# THE Gut-Immune CONNECTION

Also by Emeran Mayer, MD

*The Mind-Gut Connection*

# THE Gut-Immune CONNECTION

How Understanding the Connection Between Food and Immunity Can Help Us Regain Our Health

Emeran Mayer, MD

WITH Nell Casey

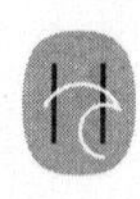

HARPER WAVE
*An Imprint of* HarperCollins*Publishers*

This book contains advice and information relating to health care. It should be used to supplement rather than replace the advice of your doctor or another trained health professional. If you know or suspect you have a health problem, it is recommended that you seek your physician's advice before embarking on any medical program or treatment. All efforts have been made to assure the accuracy of the information contained in this book as of the date of publication. This publisher and the author disclaim liability for any medical outcomes that may occur as a result of applying the methods suggested in this book.



HarperCollins books may be purchased for educational, business, or sales promotional use. For information, please email the Special Markets Department at SPsales@harpercollins.com.

FIRST EDITION

Library of Congress Cataloging-in-Publication Data has been applied for.

ISBN 978-0-06-301478-7

21 22 23 24 25 LSC 10 9 8 7 6 5 4 3 2 1

To my patients who taught me about
the importance of gut health

To the thousands of people who have
embraced *The Mind-Gut Connection* and who
have encouraged me to write this book

To Minou, who has continued to be a
persistent source of encouragement

# CONTENTS

# INTRODUCTION

In my last book, *The Mind-Gut Connection: How the Hidden Conversation within Our Bodies Impacts Our Mood, Our Choices, and Our Overall Health,* I detailed how the brain and the trillions of microbes living in our gut communicate in ways that profoundly influence our brain, gut, and well-being. I came to this perspective after three decades spent as a gastroenterologist studying brain-gut interactions in my patients.

But the world of research (and the world at large) has changed dramatically over the last five years: while microbiome science has continued to grow exponentially, and many human studies have confirmed earlier preclinical findings, our unfolding, multifaceted public health crisis has engulfed a large proportion of the US population and many countries around the globe in an epidemic of obesity and compromised metabolic health involving not only the brain, but many other organs. At the same time, as I wrote this book, the world was plunged into a pandemic in which an invisible microorganism took center stage and brought many segments of society to a sudden halt, painfully demonstrating the ingenuity and almost unlimited power of microbes.

Even though I've long held a holistic view of life, my scientific career has ultimately taken me full circle, from a reductionistic focus on the biology of brain-gut interactions back to the concept of the interconnectedness of the health of humans and of the environment, and the microbiome, with diet playing the key role in these connections. To understand its complexity and to identify a way out of our current crisis, this concept comes with a requirement for an ecological and systems view of food, health, and the environment. A "conversation" is constantly unfolding within us, influenced by our thoughts and emotions, our lifestyle, and the food we eat; the exchange between these factors occurs as a circular process in which the brain influences the gut microbial signals, which then feed back to the brain and body.

Miscommunications in this system are accompanied by a compromised regulation of the millions of immune cells located in our gut, the "gut-based immune system," resulting in a chronic, inappropriate engagement of the immune system. This chronic immune activation not only can increase gut permeability, but it can spread throughout the body, resulting in increased susceptibility to a number of chronic noncommunicable diseases and disorders—including obesity and metabolic syndrome, diabetes, and heart disease, as well as Parkinson's disease, autism spectrum disorders, depression, accelerated cognitive decline, and, ultimately, Alzheimer's disease. As we learned in the current pandemic, a compromised gut-based immune system also leaves us vulnerable to susceptibility and severity of viral pandemics like COVID-19.

Over the last decade, disorders such as these—all related to altered brain-gut-microbiome interactions—have seen such a steep increase in prevalence that they've now reached public health crisis levels. The dramatic numbers not only illustrate the scope of the problem, but also point toward the interconnectedness of many if not most NCCDs. While our health-care system, with the help of the pharmaceutical industrial complex, has been able to keep the mortality from these diseases steady, or even reduced for some of them, their overall prevalence continues to increase in younger people and in developing countries around the world.

This is where the ideas of network science and systems biology become critical. This universal conceptual approach has become essential to understand biological interactions, all the way from molecular gene networks and microbial networks to disease networks and large-scale interactions within natural ecosystems on the planet. What initially may sound like esoteric theory has in reality become an important scientific approach that offers a critical holistic understanding of health and disease. Let's take the communication between the plants that we eat and the soil they grow in—which, incidentally, has its own extensive microbiome—as an example. Microbes living in the soil interact with the root system of the plants, providing essential micronutrients and soil organic matter to their growth. The network of immune, hormone-producing, and nerve cells located in our gut wall and the gut microbiome communicate in a similar way as soil microbes interact with plant roots, even using some of the same signaling molecules. Network science is being applied to understand the interactions of soil microbes with the plants, as well as the interactions of our food, our gut microbes, and our bodies.

In addition to a poor diet, chronic stress and negative emotions impact the brain-gut-microbiome network, which is why the effects of emotional turmoil and stress mirror the negative effects of an unhealthy diet; the two seemingly unrelated but often co-occurring influences can potentiate each other. It's because the signaling molecules generated by this stress-modulated gut connectome, in particular the low-grade immune activation and many neuroactive molecules, feed back to the brain and reinforce the altered brain-gut communication. In fact, it is now becoming apparent that such circular interactions involving the gut microbiota, their metabolites, and the associated immune activation in the gut play a causative role in several chronic brain disorders, in particular depression, ASD, and Parkinson's and Alzheimer's diseases.

And so, in order to understand and ultimately overcome our current health problems of both noninfectious chronic illness and acute life-threatening pandemics, we cannot continue our futile journey from one new medication or dietary approach to the next. We must take into consideration all aspects of life and our interactions with the environment using a systems biological approach in order to return our immune system to its normal function of protecting us from invading pathogens and increase our resilience, rather than attacking our bodies.

Making sustainable changes to our diet is a crucial first step toward reestablishing health-promoting interactions between our food, our gut microbiome, and our immune system. There's a rapidly growing body of scientific evidence that shows different types of largely plant-based diets are not only *associated* with better health of the gut, the brain, and the body, but also that such diets actually play a *causative* role in better health. While this is largely demonstrated in studies exploring depression, cognitive decline, neurodegenerative diseases, and autism spectrum disorder, it can also be applied to a number of other diseases, such as coronary artery disease, fatty-liver disease, and inflammatory bowel disease.

In *The Gut-Immune Connection*, I propose a radically different approach to deciding what's best for our health, both in terms of *what* we eat and *when* we eat. First, rather than obsessing over the right amount of macronutrients we eat, I urge the reader to focus on consuming foods that support the health, diversity, and well-being of the trillions of microbes living in our gut—a consideration that is mostly lacking in the Western diet, and which continues to be neglected in the majority of fad and weight-loss diets. This change in dietary dogma means we have to eliminate ultraprocessed foods, which are packed with empty calories and chemicals but are devoid of fiber. In contrast, we have to dramatically increase microbe-targeted foods, which are poorly absorbed in our small intestine (thus providing fewer calories) and require the metabolic machinery of our gut microbiome to break them into smaller, absorbable, health-promoting molecules. These foods not only increase the diversity and richness of the gut microbiome, but they provide a large variety of fiber molecules as well as thousands of so-called polyphenols, many of which are transformed into health-promoting, anti-inflammatory signaling molecules in our gut, which after absorption into the bloodstream are distributed throughout the body.

In addition to this fundamental change in *what we eat*, recent scientific evidence has demonstrated that restricting the amount of time during which we consume food so-called time-restricted eating—has an additional beneficial effect on the rhythm by which the microbiome interacts with our gut and immune systems, leading to improved metabolic health. The most important initial step to stemming the tide of our public health crises is to curb chronic and infectious diseases not through an increasing battery of medications, but through better control of our gut-based immune and microbial systems, using the natural healing power contained in our food. This is best achieved through reconsidering the foods we consume and their relationship to our internal microbiome as well as their connection to the soil-based microbiome in which they grow. We must understand the complete microbial interconnectedness that exists not only between humans and their food, but also between farm animals and their environment and between plants and the soil. We've dramatically altered this planetary network over the past seventy-five years, and are now paying the astronomical price, in particular in the form of our current disease care system. Science is increasingly demonstrating the close connectedness between our health, what we eat, how we produce our food, and the impact of these behaviors on the planet and one another.

As pointed out by prominent scientists and organizations, it is possible to slow and even reverse the steady upward rise of illness in the United States and the world at large, even before we fully understand the universe of our gut microbes and the molecular underpinnings of each disease. We have to prevent the detrimental consequences that our food system has on the health of the planet, with a new approach based on improving the health of the gut and its microbiome and, in turn, returning the immune system to its normal, health-preserving function. While there is no question that we will conquer the current viral epidemic in the world, there will never be a vaccine to prevent and treat the worldwide epidemic of chronic noncommunicable diseases. We're in an urgent moment; consider this the ringing of our global alarm, as well as an unequivocal plan for turning things around.

# THE Gut-Immune CONNECTION

Chapter One

# AMERICA'S SILENT PUBLIC HEALTH CRISIS

When I was in medical school, in the 1970s, there was a buoyant optimism about the advances being made in the field of medicine at large. Effective treatments had been developed for many of the diseases I was studying, and several promising new interventions—such as coronary bypass surgery—were on the horizon. Even when it came to the illnesses that remained stubborn mysteries at the time—peptic ulcer disease, gastroesophageal reflux disease, inflammatory bowel disease, and various forms of cancer—there was still a tremendous sense of hope, a feeling that it would only be a matter of time before we would eradicate them, too. Unfortunately, the promise of fifty years ago has become a knot of contradictions, one that we must untangle if we're going to recalibrate and set ourselves on a path toward long-term, sustainable health and longevity.

It is true that today we are living longer than ever before in human history. In the United States and most of the developed world, the average lifespan has been extended by nearly thirty years over the last century.[1] This extraordinary progress, however, has come at a steep cost: we're also sicker than we've ever been. Over the last seventy-five years, an array of serious, seemingly unrelated chronic illnesses—cardiovascular disease, diabetes, metabolic syndrome, autoimmune disorders, cancer, chronic liver disease, and brain disorders such as depression, autism spectrum disorder, Alzheimer's disease, and Parkinson's disease—have all been steadily rising, some at astonishing rates. While living dramatically longer lives, many of us are suffering throughout them, creating a public health crisis of historic scale. Sadly, this crisis disproportionately affects minorities and populations on the lower side of the socioeconomic spectrum.

This fact has been obscured, however, by the US health-care system's practice of throwing excessive amounts of money at these illnesses, trying in vain to contain their impact. Health-care services jumped from 5 percent of our gross domestic product (GDP) in 1960 to 17.8 percent in 2019—$3.8 trillion. This number is expected to rise even higher in the years ahead.[2]

These skyrocketing medical costs are, of course, caused by a variety of factors, including the exponential growth of the medical industrial and pharmaceutical complex. Americans now, for instance, spend ten times more on prescription drugs than they did sixty years ago.[3] The costs of diagnostic tests and therapeutic medical and surgical interventions are also rising. To a significant extent, however, the unchecked increase in our medical spending is driven by the increasing prevalence of chronic diseases—and the medical establishment's immense efforts to keep death at bay, known in the formal language of my profession as "maintaining low mortality rates."

"We now have an economic system where an industry makes money off of keeping you alive but not letting you die," my friend and colleague Wayne Jonas, MD, executive director of Samueli Integrative Health Programs at UC Irvine, once succinctly summed up the situation. The remarkable increase in life expectancy achieved during the past half century has obscured the fact that the win came at an unsustainable cost, even for one of the richest countries on Earth. While we may not die from chronic illnesses as frequently as we once did, a large proportion of the population is not living into old age with any semblance of health and vitality. And we're bankrupting ourselves in the process.

This data may rightly cause you to ask, how did we get to this point?

As I'll illustrate in the coming chapters, dramatic lifestyle changes over the last seventy-five or so years are responsible for much of our illness and suffering today. Though a variety of factors have played a role in our deteriorating health—such as reduced physical exercise and sleep with increasing stress levels and exposure to a long list of chemicals and environmental toxins—the most impactful shifts have been those that have affected our food supply and our diet.

The rise of modern industrial-style agriculture has drastically changed the way we produce food as well as what and how we eat.[4] With small, family-run farms increasingly giving way to industrial farming operations, the production of our food has become more and more compromised. Industrial agriculture runs farms as factories with "inputs" such as pesticides, feed, fertilizer, and fuel and "outputs" in the form of corn, soybeans, and meat. The primary objective of these corporations is to raise profit margins by rigorously decreasing production costs and increasing yields. While food *has* become cheaper and more abundant under this system, its quality has suffered—and the health of the public (and the environment) is the collateral damage.

This relatively recent dietary shift has affected our health in myriad ways. It has changed, in some ways irreparably, the trillions of microbial organisms living throughout our gut—commonly referred to as the gut microbiome—and thus has created a chronic dysregulation in various organs and bodily systems, in particular the immune system, for the immune cells in the gut comprise 70 percent of it. As diverse and seemingly unrelated diabetes, Alzheimer's disease, and cancer may seem, there *is* a common factor that plays an important role in their coinciding surge. As I will explore in depth in the next chapter, a growing mismatch has developed between the gut microbiome—which has rapidly adapted to our changing diet—and the gut, with its much slower ability to deal with these diet-induced microbial changes. I strongly believe that this growing mismatch has disrupted the normal function of our immune system and altered our broader brain-body network, creating a stark rise in a wide range of chronic diseases.

While the overall mortality rate of infectious *and* noninfectious (noncommunicable) diseases declined rapidly in the first half of the twentieth century, the prevalence of noninfectious diseases has since reversed itself and drastically risen over the last seventy years.

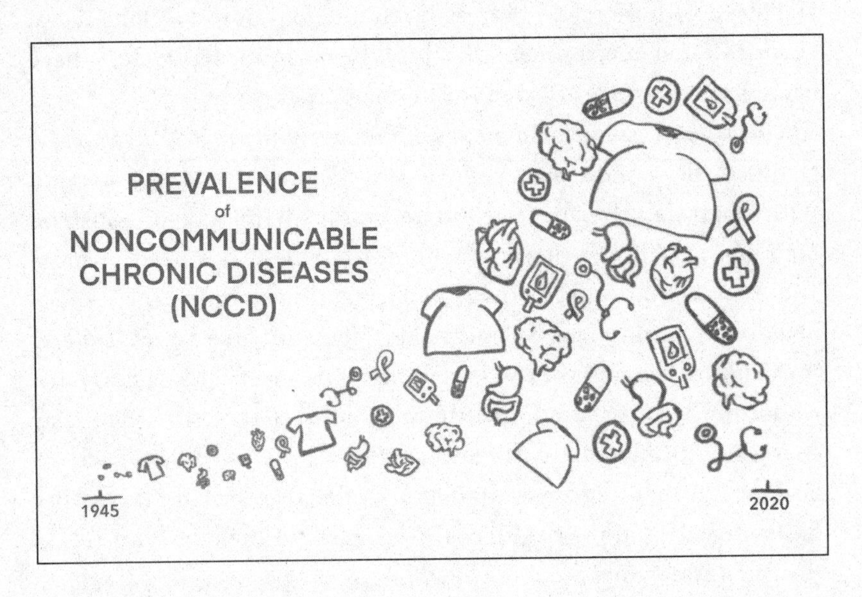


Meanwhile, most *infectious* diseases—such as tuberculosis, hepatitis A, measles, and mumps—have continued to steeply decline over the same period of time. The "theory of epidemiological transition" has attributed this shift to the decline of pestilence and famine, giving people longer lives in which the degenerative diseases have time to develop. A number of isolated spikes in infectious diseases—like AIDS, tuberculosis, Ebola, influenza, SARS, MERS, and the recent outbreak of the COVID-19 virus—have periodically occurred throughout this steady decline. However, these have not changed the overall trend: infectious diseases now account for only 4.2 percent of the burden of all diseases globally, whereas chronic diseases account for 81 percent. Moreover, noninfectious diseases today account for more than 70 percent of all deaths globally.[5] Even worse, chronic disease and pandemics often reinforce each other; we are now realizing that noninfectious diseases make us more vulnerable to certain infectious ones. For instance, COVID-19 disproportionately affects those suffering from various chronic illnesses, including obesity, diabetes, and metabolic disorders. The interrelated problems of an unhealthy diet and lower socioeconomic status are emerging as major contributors to this trend. The 2020 global pandemic was not only a tragedy in its own right, but it also highlighted the true cost of chronic disease and inequality on public health.

Thankfully, there is a way to reverse this trend.

First, however, it's important to better understand the main areas of our health that are being crucially affected by recent alterations to the gut microbiome. Of the many diet- and gut-microbiome-related chronic noninfectious diseases, I will focus on three types that play a primary role in our current health-care crisis: autoimmune and allergic disorders, obesity and metabolic syndrome (including its implications for diabetes, cancer, and cardiovascular and liver diseases), and brain disorders.

## Allergies and Autoimmune Disorders

There's an oft-cited article on allergy-related disorders that signaled a shift in the way we view chronic noninfectious diseases. Written by Jean-François Bach, MD, DSc, and published in the *New England Journal of Medicine* in 2002, the article suggested that many chronic diseases, including a group of allergy and autoimmune diseases, have been rising over the last seventy years.[6] Since its publication, a growing number of studies have offered supporting evidence for this observation. One study, for example, published in the *Scandinavian Journal of Gastroenterology*, reported that the incidence of Crohn's disease, an autoimmune disorder, more than tripled in northern Europe from the 1950s to the 1990s.[7] Another study, from researchers at the University of Gothenburg in Sweden, showed that the prevalence of asthma, hay fever, and eczema doubled in Swedish schoolchildren over the twelve-year period between 1979 and 1991.[8] Further confirmation came from researchers at Göttingen University in Germany, who looked at a population in South Lower Saxony and reported that the incidence of multiple sclerosis, also an autoimmune condition, had doubled in just under two decades, from 1969 to 1986.[9]

Several related hypotheses—the "hygiene," "old friends," and "disappearing microbiota" theories—have been proposed to explain the recent acceleration in autoimmune and allergy-related illnesses.[10] These theories all share the view that environmental factors—such as the inappropriate or excessive use of antibiotics early in life, the increased use of pesticides and chemical fertilizers in agriculture, and the rising number of children growing up in urban settings removed from nature, soil, and animals—play important roles in this shift. The hygiene hypothesis, for example, posits that in our increasingly sterile world, in which babies and small children are exposed to fewer and fewer germs and microbes from the natural environment, our immune system isn't properly trained to protect our bodies from threat. Consequently, our immune system loses the ability to discern benign substances, such as pollen or tree nuts, from hazards, such as pathogenic bacteria and viruses. As a result of this lack of discernment, the immune system either irrationally attacks the body's own cells, provoking an autoimmune disorder, or mistakenly rings the alarm bells, resulting in an allergic reaction.

The research does seem to substantiate some of these theories, at least up to a point. However, the primary focus of most studies has been to identify specific genes causing dysfunctions that increase vulnerability to autoimmune disorders and allergies, but as it turns out, no single gene has been identified as responsible for *any* major chronic diseases. Rather, a growing list of so-called vulnerability genes and altered gene networks have been identified, suggesting that a person is by nature more or less susceptible to progressively changing environmental triggers. Because our genes have not altered during the past seventy years (evolution is much slower than that), it seems almost certain that changes in our environment and lifestyle are to blame for our sudden uptick in chronic illness.

Despite the fact that the increase in such disorders first manifested itself more than half a century ago, we're still struggling mightily with them today. We've developed more-effective (and more-expensive) treatments but no straightforward cures. One need only watch the growing number of television commercials touting a legion of powerful new medications aimed at constraining an overactive immune system—with their sotto voce litany of often very serious side effects—to get a sense of the problem's magnitude. Many of these ads are for "biological drugs" or "biologics," so named because they're produced from living organisms or contain components of living organisms; think Humira, Remicade, and Rituxan—used to treat autoimmune disorders like inflammatory bowel disease, rheumatoid arthritis, and psoriasis. These medications trap signaling molecules called cytokines, which would otherwise trigger chronic inflammation and pain in the body. While the drugs have provided dramatic temporary relief to many thousands of patients, they haven't slowed the rising prevalence of these diseases.

At the same time, these treatments have created a multibillion-dollar revenue stream for the pharmaceutical industry. In large part, this is because biologic medicines cost, on average, twenty-two times as much as conventional drugs.[11] The cost for a one-year treatment with infliximab (Remicade)—prescribed for ulcerative colitis and Crohn's disease, among other ailments—is about $50,000.[12] Meanwhile, the net effect for patients has been a reduction in bothersome symptoms rather than identification or treatment of the root cause of the deviant immune system triggering such symptoms in the first place.

This shortcoming is reflected in the dramatically growing incidence of autoimmune diseases today. The American Autoimmune Related Disease Association (AARDA) estimates that fifty million Americans currently suffer from autoimmune illnesses—of which there are now more than one hundred types, including multiple sclerosis, rheumatoid arthritis, inflammatory bowel disease, and type 1 diabetes—making this group of disorders more prevalent than even cancer.[13]

And yet there is not, as there is with cancer, a consistent understanding of the factors driving its continuous increase. There is, in fact, quite a bit of bewilderment about not only the origins of these illnesses but also what exactly they are. Despite their troubling interference with the quality of life of so many people, and the prevalence of TV commercials about affected patients, 85 percent of Americans aren't able to name a single autoimmune disease. I would suggest that just as many people don't fully understand how these ailments manifest in the body or how we may be able to reduce our risk of developing such a disease.[14]

## Obesity and Metabolic Syndrome

Obesity has also played a key role in our current disease epidemic, causing a vexing rise in illness globally. In the 1960s, when the numbers of overweight and obese people slowly began to climb, the increase was barely noticed by the health-care system. Fifteen years later, when the issue finally did get attention, it was, sadly, viewed as a problem limited to minorities and to the poor in the South, revealing a racial and economic bias in the health-care system that unfortunately persists today.

Then the weight issue ballooned: between 1980 and 2013, the number of overweight and obese individuals worldwide rose from 857 million to 2.1 *billion*.[15] It became undeniable that obesity was affecting all populations and posing an unprecedented challenge to public health. Today, one in three adults and one in six children are considered obese, according to research gathered by the National Health and Nutrition Examination Survey.[16] I've observed the obesity epidemic firsthand, both in my clinical work with patients and as I travel to attend medical and scientific conferences, crisscrossing America throughout the year. As a physician, I feel a piercing sense of concern when, in airports and in line for the buffet breakfast at hotels, I see how many people appear to be on a spectrum above a normal body weight.

Though we've poured tremendous resources into research aimed at understanding the problem, we've made little progress in deciphering *why* this issue has increasingly gripped so many people over the last half century. And worse, the only current interventions that have shown long-term effectiveness have dramatic and irreversible consequences on the functioning of our digestive system. One such solution, for example, is bariatric surgery, which reduces the upper stomach in order to limit the amount a person can eat. In one form of bariatric surgery, the stomach is reshaped to the size of an egg and connected directly to the small intestine; in another, known as sleeve gastrectomy, 80 percent of the stomach is removed, leaving it about the size and shape of a banana. Another weight-loss surgery involves placing a saline-filled silicone balloon into the stomach. Still another is a drastic procedure that involves inserting a gastric fistula device (called in technical terms an "aspire assist device") that lets a person eat, then empty the contents of the stomach into a disposable bag through an artificial opening.

Not only do these surgeries illuminate the extreme medical measures we now undertake in our efforts to battle obesity, but they've also taught us that this seemingly straightforward approach—shrinking the stomach so one can't put as much food in it—is far more complicated in its effect than once thought. Such drastic interventions create all-encompassing consequences in the body, not just in the size and shape of the stomach, but also in how appetite-regulating hormones are released into the blood and reach the brain. Such operations change the composition of our gut microbes and consequently the gut's signals to the brain and the rest of the body. Even food preferences can change suddenly. There is, in other words, a consequential integral shift affecting a variety of bodily systems—hormonal, metabolic, and endocrine—even *before* weight loss begins.

Furthermore, many obese and overweight Americans have metabolic syndrome. This diagnosis is comprised of a cluster of conditions, including increased body-mass index (BMI), high blood-sugar and triglyceride levels, high blood pressure, low HDL ("good") cholesterol levels, and dyslipidemia, an asymptomatic condition in which a person's lipid profile (amount of fats in the blood) may be too high, reflecting the body's compromised ability to process sugar and fat. Most important, metabolic syndrome is not only a complication of obesity affecting the endocrine and immune systems, but also a major risk factor for chronic diseases of the liver, the heart, and even the brain.

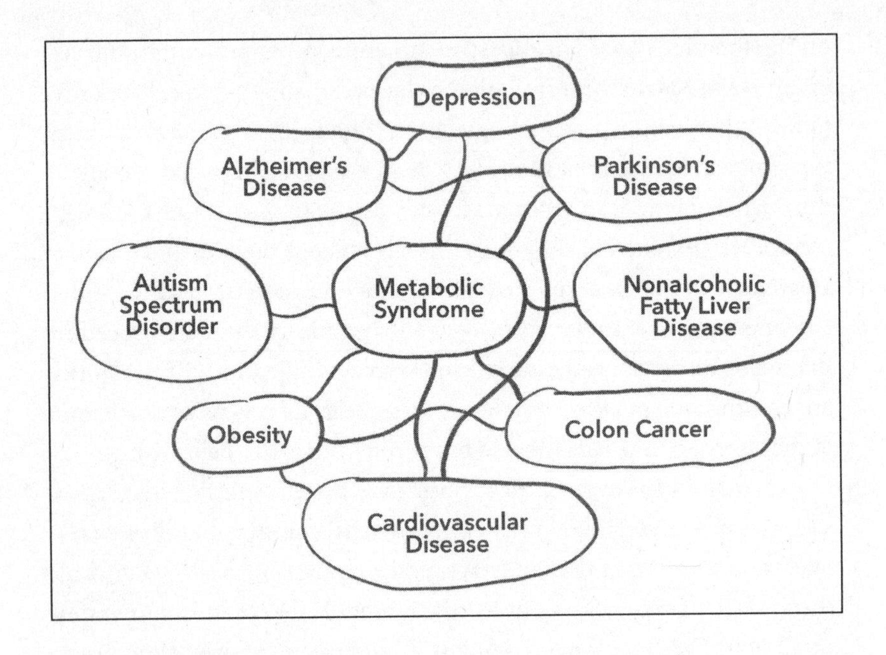


In 2018, with infectious diseases waning (before the rise of COVID), one study declared metabolic syndrome "the new major health hazard of the modern world."[17]

Some experts believe that we're only at the beginning of this trend. As Dr. Walter Willett, professor of epidemiology and nutrition at Harvard University, explained to me, "This epidemic of obesity and insulin resistance takes thirty, forty, fifty years to play out before we see all of the consequences. It's somewhat like climate change. You don't see all the implications right away, but we can see the path is leading to devastation in terms of health." Sadly, like obesity, metabolic syndrome is no longer limited to the developed world. In China, for example, the prevalence of overweight and obese people increased from 20 to 29 percent of the population from 1992 to 2002, and by 2017, the prevalence of metabolic syndrome had jumped to 15.5 percent.[18]

As a result of these soaring rates of metabolic syndrome, cardiovascular diseases—including high blood pressure, coronary heart disease, heart attack, stroke, congestive heart failure, and atrial fibrillation—have also steadily increased, because metabolic syndrome is a major risk factor for these diseases. In 2011, the American Heart Association predicted that up to 40 percent of the US population would suffer from some form of cardiovascular disease by 2030, but we reached this benchmark in 2015, in four years instead of nineteen.[19] In 2015, 96 million Americans suffered from high blood pressure and almost 17 million from coronary heart disease. As if this wasn't discouraging enough, this ominous trend is expected to increase to 45 percent in another fifteen years—another prediction we may well beat.

Similar to these numbers is the unreasonably high cost of prescription medications, surgeries, and hospital stays to keep patients with metabolic syndrome alive. In 2016, this price tag in the US was $555 billion; it is expected to exceed $1 *trillion* by 2035.[20]

It seems no organ can escape the impact of a dysregulated metabolic system. It's estimated that 75 percent of patients who are overweight and 90 to 95 percent of patients who are morbidly obese are afflicted by nonalcoholic fatty-liver disease, or NAFLD, a serious condition that can lead to cirrhosis, liver cancer, and liver failure. It is the leading cause of liver disease in the United States and one of the major indications for liver-transplant surgery.[21] Obesity and metabolic syndrome are also significant risk factors for several types of cancer, including those of the colon and rectum, the fourth most common form of cancer in the US. According to the National Cancer Institute, obese individuals, especially men, are about thirty times more likely to develop colorectal cancer than people of normal body weight.[22]

## Brain Disorders

Several psychiatric, cognitive, and neurodegenerative disorders—including Alzheimer's and Parkinson's diseases, autism spectrum disorder (ASD), depression, and anxiety disorders—have also affected a significantly higher proportion of Americans over the last half century. Even though this increase hasn't been as dramatic as that of obesity and metabolic syndrome, the trend is still quite striking. In the last twenty years, neurodegenerative disorders have ticked upward. In 2017, an estimated fifty million people were suffering from

Alzheimer's disease worldwide,[23] and that number is expected to double every twenty years into the foreseeable future.

While our increase in life expectancy certainly plays a role in such estimates, we have evidence that a variety of other factors, including metabolic syndrome, are also instrumental in the development of cognitive dysfunction. Sadly, we've almost come to accept cognitive decline with age as the norm, just as we've accepted the implicit messages of the pharmaceutical industry that the recent rise in many of our chronic diseases is simply a by-product of aging. In actuality, and as many fully functioning nonagenarians demonstrate, the human brain (as well as the body) has the potential to function fully into our nineties without any medical interventions.

Other chronic diseases are also increasing rapidly. In the 2016 world, 6.1 million people globally had Parkinson's disease,[24] whereas today, more than 10 million are living with it.[25] And, as I'll explore further in chapter 4, developmental disorders like ASD have nearly tripled in frequency, from 1 in 166 children in 2004 to 1 in 59 children in 2018.[26] Depression, too, is being identified in increasing numbers, although it is slightly more complex in its presentation, and thus the change in its prevalence is harder to assess, particularly given that it is not a homogeneous condition. It can, for example, occur alongside other illnesses, such as Parkinson's and Alzheimer's. Still, in 2017, about 160 million people suffered from major depressive disorder, young people being the highest-risk group.[27] A Blue Cross Blue Shield report showed that in 2016, in fact, 2.6 percent of youths between twelve and seventeen years of age were diagnosed with major depression—a 63 percent increase from 2013. Among young adults of ages eighteen to thirty-four, the report showed a 47 percent increase. Even more unsettling, a recent study predicts that the number of younger people diagnosed with depression will surpass even the number of cardiovascular disorders in adults by 2030. In addition, suicide, considered a proxy measure for the prevalence of depression, has become a leading cause of death for young people in the United States—and is the only one among the top ten causes that's continuing to rise in numbers. Even though we haven't been able to find a solution or even consistently successful treatments for depression, the pharmaceutical companies are enjoying tremendous profits: mental-health drugs bring in an estimated $80 billion a year worldwide.

## The Common Link

Over the last seventy-five years, many chronic illnesses have been researched and treated as distinct issues, each one seen as its own discrete epidemic and therefore dealt with by specialized physicians and researchers. Nevertheless, in that time, our modern health-care system hasn't been able to create an effective strategy for halting the steady upward trajectory of *any* of them.

However, if viewed through the understanding that these noninfectious diseases have *simultaneously* overtaken huge swaths of the population, striking parallels begin to emerge. Many of these illnesses, for example, have shown a similar lag between their increase in the developed world and in the developing world, following the timeline of accelerated industrialization. Approximately ten to thirty years after the increase of these illnesses in the Western world, the same rise has been echoed in less developed countries, *after* they've adopted many aspects of our modern diet and lifestyle. Take, for example, the inflammatory bowel diseases ulcerative colitis and Crohn's disease. At the turn of the twenty-first century, these have become global diseases with accelerating incidence in newly industrialized countries throughout Asia, Africa, and South America.[28]

In addition, with allergy-related illness, autoimmunity, obesity, metabolic syndrome, colon cancer, and depression, we have seen a shift downward in age of onset, as they begin affecting an increasingly younger population. This suggests that our recent dietary alterations are also affecting subsequent generations. For example, while the incidence of colon cancer has decreased in older men and women, the rates among younger men and women are climbing. Between 2006 and 2015, the average annual increase in colon cancer among men under the age of fifty has been 3.5 percent.[29] The way the health-care system has responded to this worrisome trend is emblematic of its reductionist and shortsighted approach. The American Cancer Society Guideline for Colorectal Cancer Screening now advises that people at average risk for colon cancer—without family history or other well-known risk factors—start regular screenings at age forty-five as opposed to the prior recommendation of sixty.[30] And yet when I recently attended a lecture about the guidelines for colon-cancer screening, I asked the presenter if diet and childhood obesity might play a role in the shift and, if so, whether dietary recommendations might be given as a preventive measure. She agreed that this was a plausible explanation but that offering dietary recommendations at screenings is currently not part of the established practice. Besides, she went on, with many gastroenterologists performing such a large number of colonoscopies each day, there wouldn't be enough time to explore patients' eating habits or provide guidelines for a beneficial diet. It's always amazing to me when simple changes—particularly those that could potentially have a large-scale impact but don't fit into the traditional disease model—are so easily dismissed.

Furthermore, despite these illnesses' widely varying manifestations, almost all of them can be traced to a dysregulation of the immune system. This plays out in one of two ways. With autoimmune and allergy-related disorders, the immune system overreacts to benign environmental stimuli or to the body's own cells, whereas in metabolic and related diseases there's a chronic, unprovoked engagement of the gut-based immune system that can affect *all* organs in the body, including the heart, liver, large intestine, fat tissue, and even the brain. (The latter tendency of the immune system to overreact—also referred to as a cytokine storm—may be related to the increased vulnerability of COVID-19 patients with chronic, noninfectious diseases for developing more severe symptoms and complications.)[31]

Although there's good evidence to suggest that alterations of the immune system predisposing a person to autoimmune and allergic diseases are programmed during the first three years of life, a growing body of evidence suggests that the Western diet also plays a key role in the development of these and other illnesses comprising our current public health crisis. Such a diet may lead to *metabolic endotoxemia*, a low-grade systemic immune activation that carries inflammatory mediators—including messengers such as the cytokines, a family of signaling molecules secreted by various cells of the immune system—throughout the body and brain.[32] Because 70 percent of our immune system is located in the gut's walls, it's in a powerful position to spread this inflammatory message throughout the body, and depending on a person's genetic variability, the cytokines affect any of various organs along the way.

It may seem counterintuitive to deduce that our digestive tract, long viewed as an organ primarily concerned with the absorption of nutrients and storage and elimination of waste, would be a main actor in this radical unfolding health drama. But a rapidly growing body of research from the past two decades, strongly influenced by the expanding discipline of systems biology, leads us to this conclusion in extraordinary ways. As I'll explore in the forthcoming chapters, recent discoveries made about the gut microbiome and its relationship to the brain and *all* systems in the body—including the immune system—is one of the most exciting steps forward in understanding how we might begin to halt and even reverse our formidable public health crisis.

**Chapter Two**

# A DEEPER CONNECTION

Recently, science has both returned to *and* arrived at a vastly different understanding of the body and our health. This view is one that accounts for complexity, communication, and the interrelatedness of bodily systems, as well as making clear how so many seemingly disparate illnesses, all rising at precipitous rates over the last seventy-five years, are connected.

I regard this new holistic view as a return to a previous way of thinking. The notion of the interconnectedness of bodily systems can be found in Ayurvedic texts dating from five thousand years ago. This concept was also embraced in traditional Chinese medicine and Hippocratic medicine (which was based on the natural philosophy of the Greeks). In ancient times, there was a recognition that our health was determined by an intricate relationship among mind, organs, spirit, environment, and even the universe.[1] More succinctly, as the Greek philosopher Aristotle wrote more than two millennia ago, "The whole is something besides the parts."

This belief began to change in the seventeenth century when French philosopher René Descartes introduced the principle of reductionism in his philosophical and autobiographical treatise, *Discourse on the Method of Rightly Conducting One's Reason and Seeking Truth in the*

*Sciences*. This is where the famous phrase "I think, therefore I am" originates. With reductionism, Descartes proposed that we should analyze complex situations by dissecting them, examining them in manageable pieces, and then reassessing the whole based on the behavior of the parts.[2]

Later, when Descartes proposed mind-body dualism—treating body and brain as completely separate entities—he brought reductionism to bear on the body.[3] To reconcile the mind-body conflict, he proposed that doctors and scientists concern themselves with only the body, while the brain and mind should fall under the dominion of the Church. Descartes's views altered not only philosophy, but biology, too. The medical world adopted reductionism and dualism, and physicians began diagnosing and treating the body based on the premise that it is made up of discrete parts, each one functioning separately. Every living thing, doctors came to believe, was made of mechanisms that operated as regularly and predictably as the gears of a clock. Despite this centuries-long detour, science—if not the medical establishment—is slowly circling back to the ancient wisdom traditions' view of the body as a complex interplay of interrelated systems.

Now, of course, we have a better understanding of the biology that underpins these connections. The introduction of network science was one of the biggest influences in shifting our perspective back to the unified from the isolated, while also moving it forward with scientific corroboration.[4] Network science studies the interplay among individual elements in complex networks using such methods as graph theory, statistical mechanics, and data mining to create predictive models. Developed in the 1930s, since then it has dramatically accelerated and expanded through diverse scientific fields, ranging from the social sciences to ecology to the global economy. As a result, we now view many collections of seemingly unrelated elements as systems made up of closely interconnected parts with predictable patterns but often unpredictable outcomes.

"Think people, the stock market, genes, neurons, molecules in a cell. The interactions are what counts," said my friend and colleague Olaf Sporns, provost professor in psychological and brain sciences at

Indiana University as well as co-director of the university's Network Science Institute. "We need a science that deals with the complexity of systems like that and makes them into mathematical form, an alliance with computational methods. *That* is network science."

While network science has been applied to natural, social, and technological systems for several decades, more recently it's been applied to complex biological systems, allowing us to see the human body as an intricately drawn map of interconnections devised by mathematical sequencing.

At the same time, an approach called systems biology—first developed in the 1950s and fully adapted to modern biology some twenty years ago—has also gained traction. This began with the early promise of mapping the human genome, an endeavor that, many believed, would quickly revolutionize medicine. Bill Clinton, at the time, referred to human genetic code as "the language in which God created life." Unfortunately, many billions of dollars later, the Human Genome Project has not yet come up with practical diagnostics and treatments for the most common diseases. Still, systems biology gained traction in the medical field, in particular microbiome science, to which it offered a more complex theoretical and computational method able to capitalize on the exponential growth of supercomputers to process huge biological data sets. Scientists thereby seek to understand the body and brain by viewing their different types of cells, molecules, and microbes as comprising one system.

Systems biology gave momentum to the paradigm shift within science from specialization to interconnection. Each of the domains within systems biology is referred to with the suffixes *-ome* and *-omics*; genomics was the first such domain. From there, what I like to call the "-omics revolution" followed, with new fields emerging one after the other. Epigenomics studies the environmental influences acting on all of our genes to modify their expression (whereas epi*genetics* looks at the effects of the environment on specific genes).[5] Transcriptomics studies the set of RNA molecules expressed by genes important for the synthesis of molecules; metabolomics, the large number of signaling molecules generated by gene expression;[6] proteomics analyzes the complete set of proteins expressed by a specific cell or organism;[7] and microbiome science studies the complete set of microorganisms living in the intestine, as well as their genetic makeup.[8] What the ancient wisdom traditions understood based on centuries of astute observation, systems biology has since rediscovered by running the numbers; each of these domains interacts with and modifies the others, creating a huge interdependent, multiscale network in the body.

More recently, systems biology has been applied to two of the most intricate systems in our bodies—the *brain connectome* and the *gut connectome*. Olaf Sporns largely pioneered the field of brain connectomics, mapping the entire set of connections within the brain,[9] an intricate web of billions of neurons interconnected by trillions of synapses, a knot of fibers that if laid end to end would reach halfway to the moon. By mathematically analyzing these systems, Sporns was able to map the connections within the brain, which has led to a completely different understanding of its structure and function and, consequently, the characterization of its diseases. At the other end of the brain-gut axis, Rodger Liddle, a gut neurobiologist and professor of medicine at Duke University, proposed the notion of the gut connectome in 2015.[10]

Liddle's suggested network consists primarily of the nerve cells of the enteric nervous system, which can control a range of gastrointestinal processes independently of the central nervous system and therefore is often called the little brain in the gut. He also included other types of nerve cells, the supporting cells (collectively known as the glia) and the hormone-containing cells. I propose to extend this network in order to encompass the intricate interplay between the gut's immune system and the various other cells in the gut—and the critical role this communication plays in our health. Consequently, the *gut connectome*—or simply the gut, as I will interchangeably refer to it throughout—encompasses not only the gut's nervous system but also its endocrine and immune systems, which together regulate metabolism and the intake of food and defend the body against pathogens. Please note that when I refer to the gut connectome or the gut, I am referring to the organ, whereas when I refer to the gut microbiome, I am speaking of the trillions of microorganisms living in it.

* * *

As viewed by systems biology, the gut and its microbiome are a key to understanding the diseases of our current public health crisis, because the science has shown that the gut is the central link in the body's communications network connecting its various organ systems. In order to explain how the gut serves as this essential link, let me step back for a moment to describe network science in a bit more detail. In the vernacular of this discipline, complex networks are described in terms of *nodes* (the individual elements in the network) and *edges* (the connections or pathways among the nodes).

Here's a simpler way to think of it: there's a common saying, "All roads lead to Rome." In the ancient Roman empire, all roads (edges) eventually led to Rome, the most important node in that particular network. Like modern big cities, Rome was distinguished not only by its physical connectedness but also by its impact on the entire country—or, as a network scientist would say, its *centrality*. Measures of centrality indicate the importance of a node's influence on communication and information flow within a large network. Two other terms name the most fundamental attributes of each node: its *degree*, or number of pathways attached to it, and its *strength*, its total level of involvement in the network. The fact that Rome played such a crucial role in the ancient empire and that it had such a high number of connections with other nodes (cities) means that it was a *hub* in this network.[11]

Now try for a minute to imagine the body, *your* body, as a network in which all of your organs are the nodes. Some are more important than others for overall function; these are the hubs. The pathways, or edges, are the different ways that the biological systems communicate with one another. Some of these pathways are hardwired, such as nerve bundles and the vascular system, while others are highly dynamic communication systems—circulating immune cells, the myriad circulating molecules (hormones, inflammatory molecules, metabolites), and even the blood cells.

One of the most remarkable concepts of network science is the *scalability* of systems, meaning that even though networks are made up of such different entities as genes, molecules, cells, organs, and even people, the basic properties of networks, their behaviors and responses, are determined by the same mathematical rules. From genetic and molecular networks to social networks of people—all operate in an interrelated way. Interactions occur up and down the scale, from the most basic biological exchange to the most complex social systems and all the way back down again. For instance, alterations in gut microbial networks caused by diet can alter brain networks, leading to behavioral changes in social interactions, which then affect brain networks again, ultimately resulting in more changes at the level of gene expression in microbiome networks.

It follows, then, that the various systems of our bodies, from small to large—from the gut connectome to the brain connectome to the brain-body network—are not only in constant communication with one another, but are also constantly affecting one another. The organs with more connections to other organs are the hubs, each made up of "small-world networks," direct routes connecting each of them to all of the others. The structure of our organ network—the connectivity of nodes and the number of edges—is influenced by other networks operating on a different scale, such as our individual gene networks.

Though research on this new holistic view of the body is still developing, there's no doubt in my mind that the brain and gut—linked by thick nerve cables and myriad signaling molecules circulating in our blood vessels that transfer information in both directions—are the most important hubs in the organ network of the body. Changes in these two major hubs create a ripple effect throughout the body. Here's an example of how this works in the outside world: A blizzard in Chicago disrupts the network hub O'Hare airport. International flights get canceled, then domestic flights get canceled, and eventually people are stranded. The ripple effect from disruption at the hub ultimately interrupts or shuts down the entire network.

Today we are seeing the effect of our modern "blizzard" of lifestyle changes, which have created disruptions at the hubs of our brain-body network that are likewise interrupting or shutting down essential transactions. Based on growing scientific evidence, I believe that alterations in our brain-body network are responsible for the various diseases of our health crisis. These alterations have occurred because our bodily systems have been exposed to continual challenges since the beginning of industrialization, with a dramatic acceleration over the last seventy-five years. These challenges include pollution of the air, soil, and water; toxic chemical exposure; urbanization; the overuse of antibiotics and other medications; chronic stress; and most important, our increasingly unhealthy diet. All of these influences affect our gut microbiome and thus the whole of our health.

These perturbations have profoundly changed the ancient, beneficial relationship between the microbiome and the gut connectome.[12] Typically, the interaction between the two allows for some amount of mismatch and disruption. That is, the two can adapt in partnership to a range of challenges, such as benign infections, short-term use of antibiotics, and gradual changes in diet. The microbiome adapts with much greater flexibility than the rest of the body to the ever-shifting world around it, but the sustained pressure from modern lifestyle has prevented it staying in synch with the biology of the gut. This growing mismatch has resulted in a threat to the long-standing symbiotic relationship between the gut and its microbiome.[13]

These long-term disruptions have also altered the edges, the pathways connecting our organs, or nodes. They have changed the complex molecular language of the body, as well as the microbial metabolites generated in the gut. These changes in interorgan communication—especially the crosstalk among the brain, liver, heart, and intestines—have compromised the organs' functions. The result has been a structural and functional remodeling of the *entire* brain-body network,[14] which, in my opinion, explains the wide range of illnesses that have risen simultaneously in recent decades. If we are to make headway with our most pressing public health issues, we must begin by addressing the growing mismatch of the gut, its immune system, and its critical interactions with the microbiome.

* * *

There are many reasons—other than the fact that as a gastroenterologist I have spent most of my career studying and treating gut-related disorders—that I place the gut in such a prominent position over other vital organs, such as the kidneys, heart, and lungs, in the brain-body network. The bidirectional communication between the brain and the gut, which seems counterintuitive at first glance, was actually established deep in evolutionary history, and proof of it has since only grown stronger. This relationship dates back some six hundred million years, when the earliest multicellular organisms appeared in the planet's oceans. These tiny animals, called hydras, were not much more than floating digestive tubes with nerve nets wrapped around them. The only function of their early enteric nervous system, which technically could be considered the first brain, was to assure the proper functioning of this primitive gut, moving food from one end, the mouth, through the tube, extracting the nutrients and distributing them to the rest of the body (mainly tentacles), and then expelling the residue at the other end. It's astonishing that this earliest gut, with its close connectivity between nerve cells and smooth muscle cells, has been conserved in evolution over hundreds of millions of years, and in principle it is still shared by almost all animals on this planet, from bees to fish to elephants to humans.

Communications within the gut grew more complex when some microbes from the ocean decided to settle inside these archetypal guts, developing close communication with the nerve cells of this first brain approximately five hundred million years ago. As evolution progressed, the unique design of this original gut connectome was largely preserved while animals gradually developed a *second* brain—the one we call the brain or central nervous system (CNS) today. The signaling molecules that our first brain had developed were then incorporated into this new brain, creating a common language between it, the gut, and its microbes. This formed the basis for unique interactions within the gut-brain network that remain functional to this day. Some of these interactions form a small-world network within the gut connectome, primarily concerned with the optimal functioning of this organ (peristalsis, secretion, blood flow, food sensing). But now this small-world network is also connected to the brain via long-distance connections, by which the CNS closely monitors the gut connectome's activities and coordinates them with the overall needs of the body. This two-way conversation tells us when we're hungry or full and plays an important role in regulating our mood and well-being.

The earliest gut and nervous systems were (and are) so tightly interwoven in the hydra, it was as if they were one, and they've preserved this profound connection throughout the evolution of later animals, even as they moved farther apart in the body. Our other organs didn't develop until later and so weren't able to establish the same intimate bond, reinforcing the idea that the gut and the brain are the two major hubs in the organ network.

In addition, research has revealed that, next to the brain, the gut is the most complex organ in our bodies.[15] It has its own nervous system—sometimes called the second brain,[16] even though it's actually our *first*—as well as its own immune system and hormone-producing endocrine system. In fact, these enteroendocrine cells, which make the chemical messengers that regulate our food intake and well-being, constitute the biggest endocrine organ in our body. These cells are all part of the gut connectome and can release hundreds of different signaling molecules into the bloodstream and the gut lumen (essentially the inside of the gut, where the microbes live and where food passes by), as well as onto nerve endings within the gut wall. Most of these nerve endings are sensors of the vagus nerve, which conveys messages between the gut and the brain.

Perhaps most important, more than 70 percent of our immune cells are located in the gut wall. From there, they can either travel to other parts of the body on their own or communicate with the rest of the body by inflammatory molecules released into the bloodstream.[17] Immune, endocrine, and nerve cells are sandwiched in between the layers that make up the gut wall, and they are separated only by a thin coating of mucus from the trillions of microbes that make up your microbiome.[18] Certain immune cells, called dendritic cells, send their tentacles all the way into the mucus layer, putting them even closer to the microbes' influence. Any change in the mucus layer, either in its chemical makeup or in its physical thickness, can therefore have major effects on the exposure of the gut microbes to these sentinels of the immune system.

While the specific functions of the nervous, endocrine, and immune systems in the gut have been studied in great detail, it has only recently become clear that their interactions with one another, the brain, the gut microbes, and the food that we eat can best be understood when all of these elements are viewed as part of an interconnected whole, a system. When these interactions are synchronized in a harmonious fashion, the gut is healthy, but when there is miscommunication, it affects the normal function of the gut, and, as we have come to understand through systems biology, the effects can ripple throughout the entire body.

**Chapter Three**

# THE EMERGING VIEW OF A HEALTHY GUT MICROBIOME

One of the most remarkable experiences of my professional life has been to witness the recent explosion of public interest in the gut microbiome. For much of my forty-year career as a gastroenterologist focusing on the interaction between the brain and the gut, most of what I researched was of little interest to my colleagues and was often misperceived by the lay public as an attempt to explain gut symptoms in irritable-bowel syndrome (IBS) patients through psychology. Over the last decade, however, the gut and the microbes it houses have been recognized as affecting a wide range of activities and conditions, from athletic performance to corporate leadership and from depression to Alzheimer's. The microbiome has moved from obscurity to center stage in the scientific world and among the general public. Now, it seems, nearly everyone can speak fluently about their gut microbiome. And yet gut health, as it's promoted by the media and understood by the health-conscious public, is still a vague concept.

What exactly a healthy gut should look like is not only a difficult thing to pin down, but also our culture's enthusiastic embrace of the microbiome has created a somewhat superficial, at times distorted, interpretation of the role it plays in our lives. What is most perplexing about this misunderstanding is that, while it's overblown—with unverifiable promises that improving your gut health will boost your energy, remove your brain fog, or miraculously make you lose weight—it also overlooks the more profound and consequential news that gut health is connected to a wide variety of illnesses affecting millions of people.

Part of the confusion about gut health stems from the misperception that we're looking for a fixed, ideal state of the microbiome and that once this state is achieved, we'll reach a kind of gut-and-health utopia. This simply isn't how the microbiome operates. Despite the fact that the gut and microbiome work together in intimate partnership, there's a distinct difference between them. While the gut connectome remains relatively stable, the microbiome is constantly changing. The microbe population rapidly adjusts to changes in the gut environment and is so adaptive that it serves as a vivid reflection of the shifting world around it, in particular the diet we feed it. Though all of our organs are, to some degree, adapting to their environment, no other system in our body modifies as fast as the microbiome.[1]

Our human biology is determined by twenty thousand genes optimized and selected over millions of years of evolution. It has been estimated that some of these genes can adapt to new environmental conditions, including dietary changes, within a timeframe of fifteen to twenty thousand years.[2] Our gut microbiome is made up of an estimated four hundred thousand genes with much faster generational turnover, allowing for twenty times the ability to acclimate to changing environments, even those that the microbes have never encountered before.[3] However, despite their differences in adaptability, the slow coevolution (beginning with the hydra) of gut and microbiome has resulted in a symbiotic match, allowing humans to enjoy optimal health while living in a variety of places and eating a variety of diets over hundreds of thousands of years.

## The Healthy Gut Connectome

What makes a healthy gut? There are three closely linked components that determine gut health. The gut's endocrine system produces hormones that regulate food intake and metabolism (among other functions). Its immune system protects against pathogens and maintains self-tolerance, so the body recognizes self-produced antigens as nonthreatening while still mounting a suitable response to foreign substances. The gut's enteric nervous system regulates its peristaltic contractions and the secretion and absorption of fluids.

From a *metabolism* viewpoint, we can define *gut health* as a state in which the hormone-producing cells produce enough of the ones that make us feel hungry when the body needs energy, and enough of the satiety-signaling molecules after a meal in order to tell the brain that it's time to stop eating. If this element of the gut connectome isn't working properly, a never-satisfied feeling of hunger results, causing one to eat beyond one's metabolic needs, resulting in weight gain and a predisposition to type 2 diabetes.

From an *immune-system* viewpoint, a healthy gut is one in which the gut-based immune-system cells are isolated from the gut microbes by a barrier made up of a layer of tightly connected cells (the gut epithelium), as well as a protective mucus layer. This double defense is to prevent chronic immune system activation by the gut contents, in particular its microbes. As a growing number of studies have shown, this intestinal barrier can be compromised by an unhealthy diet with too little fiber and too much sugar, fat, emulsifiers, artificial sweeteners, and other additives. If the gut microbes no longer can feed on their main staple—a rich variety of dietary fiber—they turn their voracious appetite on sugarlike molecules called glycans or polysaccharides, which make up the mucus layer. Such a fiber-poor, diet-induced loss of the protective mucus layer brings the dendritic cells' tentacles in closer contact to the microbes, spurring them to report a potential threat to the deeper parts of the gut-based immune system.[4] When this happens, inflammatory molecules are released, loosening the tight junctions between the cells of the epithelial layer

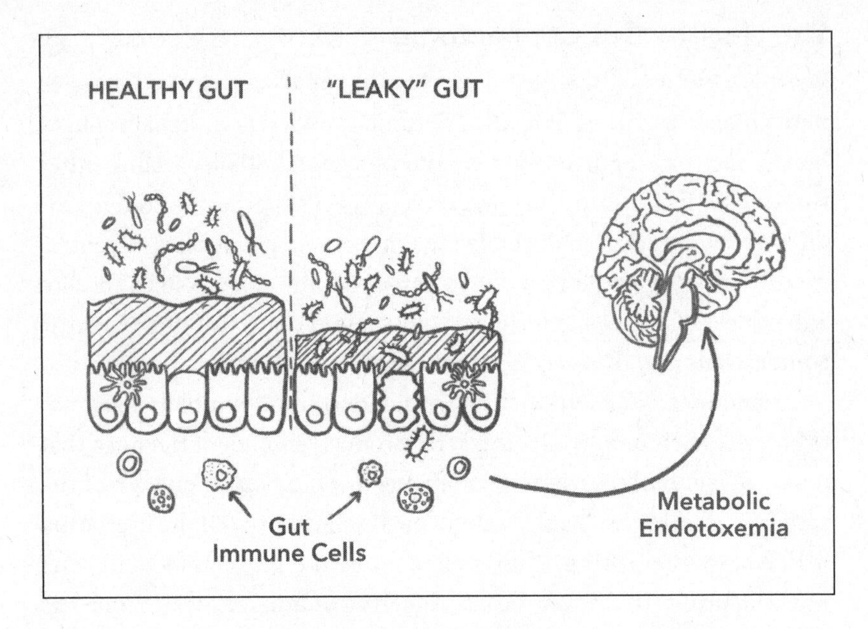


and permitting certain microbes to come into direct contact with the millions of interconnected gut immune cells. This condition is now popularly known as leaky gut.

From an *enteric-nervous-system* viewpoint, a healthy gut can be defined by the integrated activity of millions of nerve cells properly regulating its contractions and secretions. These nerve networks coordinate the parts of the gut to contract in patterns that optimize digestion and move the contents gradually from stomach to large intestine. When the gut is completely empty, its nervous system induces rhythmically recurring high-amplitude contractions that slowly move throughout the entire gastrointestinal (GI) tract, the so-called migrating motor complex. These contractions move food residues, secretions, and microbes from the upper GI tract with a low density of microbes into the densely populated large intestine. If this gut brain isn't working properly, stomach pain, irregularity, or functional GI disorders, like irritable-bowel syndrome or small intestinal bacterial overgrowth (SIBO), may result.

In a healthy gut, the endocrine system, immune system, and enteric nervous system operate seamlessly to provide us with nutrition, regulate our food intake, and protect us from life-threatening intestinal infections. In a healthy gut, these vital functions occur without requiring any of our attention, completely outside of our conscious awareness.

## The Healthy Microbiome

While defining a healthy gut is relatively straightforward, determining what constitutes a healthy gut microbiome is more complicated. Though sometimes mistakenly referred to as a microbial organ, the microbiome is far more flexible than our organs and therefore can't be regarded in the same static way that we view the liver, kidneys, or heart. But that's not the only reason it's hard to define a healthy microbiome. Despite the recent surge of interest and inquiry, we're still in the nascent stages of fully comprehending it.

Serious microbiome science was jump-started a little bit more than a decade ago by the National Institutes of Health (NIH) with the establishment of the Common Fund Human Microbiome Project (HMP) in 2008 to comprehensively characterize the human microbiome. Six years later, after an impressive accumulation of scientific data, the NIH launched a second phase of the project for a more complete profile of the human microbiota and its impact on human health and disease. Although this research is still in its infancy, the first phase of the project has generated an optimism like that surrounding the completion of the first survey of the human genome in 2000.[5] The human genome proved to be far more complex than we'd anticipated, and new research has unfurled at an impressive pace, revealing the processes behind gene regulation and expression, including the essential role of epigenetics. Although the mapping of the genome didn't provide the immediate answers we'd hoped for, it did pave the way for the emergence of fundamental changes in medicine, including gene therapy, genetic engineering, and genetic testing, such as the DNA analysis offered by the company 23andMe. In many ways, the evolving field of genomics has already revolutionized medicine.

Likewise, the relatively new discovery of the microbiome has generated a tempest of excitement, and many in the medical field (and the media) have rushed to the same conclusion—that we've found immediate explanations and treatments for many medically unexplained symptoms of chronic illness. In reality, however, we're only at the beginning of understanding this sophisticated and complex system, and as with the human genome, we're still unraveling its full meaning.

Early in our understanding of the microbiome, researchers reported the existence of a "core" microbiome—certain groups of microbes universally present in healthy individuals around the world—and assumed that a change in these microbial groups would indicate an unhealthy microbiome. However, newer studies have used technology that lets us study the microbiome at a much finer resolution of types than the initial sequencing methods did, at the level of subspecies and strains.* These studies have revealed that there is an incredible amount of

---

* Like all living things on the planet, microbes are classified by biologists into different taxa—types or categories. This taxonomic (literally "orderly arrangement") system allows us to understand how closely or distantly various organisms are related to one another. There are eight major levels of taxa, from least specific to most specific. At the top of the hierarchy is the broadest taxon, the domain, of which there are just three; just below it is the next-broadest one, the kingdom, and after that the phylum. At the bottom of the hierarchy is literally the most specific category, the species. Each phylum includes a wide range of organisms that have evolved over hundred of millions of years, while a species is a subset of the organisms within that phylum that are most closely related to one another. Just above species is the genus, a group of related species.

To illustrate, let's look at the example of humans and great apes: Humans belong to the animal kingdom and the phylum Chordata (all animals with a spinal cord). We are in the class of mammals, the order of primates, and the family of hominids. Our genus is *Homo* ("human" in Latin), and our species is *sapiens* (Latin for "wise"; the jury is still out on that point). In the same way that we use the term *Homo sapiens* for human beings, we use the term *Bacteroides fragilis* for a particular kind of bacteria of the phylum Bacteroidetes, the kingdom Eubacteria, and the domain Bacteria. Our closest living relatives are the great apes—gorillas, chimpanzees, and orangutans. They and humans make up the Hominidae family. Like humans, each of the great-ape types has its own genus. Gorillas are classified in the eponymous genus *Gorilla*, which contains two species; chimpanzees belong to the genus *Pan*, which also consists of two species; and orangutans make up the genus *Pango*, also with two species. Humans are the only species in the *Homo* genus, although a number of other, now-extinct species existed at one time—including, most recently, *Homo neanderthalensis*, or Neanderthals.

The two common bacteria types *Prevotella* and *Bacteroides* belong to two separate genera (the plural of genus), meaning that they differ from the standpoint of biological classification to the same degree that we humans differ from our pongid (ape) cousins.

variation in the makeup of individual microbiomes, and so the idea that certain species could be used to classify one as healthy is erroneous, and the concept has since been abandoned by most experts. However, much of the media and some members of the scientific community still adhere to the notion that there's enough continuity from one microbiome to the next to claim that specific imbalances in microbial populations are associated with poor gut health and can be considered diagnostic for certain diseases—such as Parkinson's or Alzheimer's disease, irritable-bowel syndrome, or inflammatory bowel disease. One example of this persistent misconception is reflected in a recent study published in *Nature Communications* proposing that researchers were able to differentiate a healthy person from one with a chronic disease simply by looking at the prevalence of a group of certain microbes in the gut microbiome. However, the researchers didn't look at the specific function of the microbes, the microbial groups were identified only to the species (and not to the strain) level, and the predictions were only 73.7 percent accurate.

We now know that in industrialized societies, as few as 10 percent of microbial strains may be shared among healthy individuals. This discovery has led researchers to explore the idea of a *functional* core as a way of characterizing a healthy microbiome.[6] After all, it's not the microbes themselves, but the compounds and signaling molecules they produce that carry their messages, interact with one another, and communicate with the gut, the immune system, the brain, and the rest of the body. Through metatranscriptomics and metabolomics—scientific disciplines by which we measure which genes are expressed and molecules produced within the microbiome—we now know that the widely varying gut microbial communities in healthy individuals are able to produce a similar set of metabolites and signaling molecules. Ultimately it is these molecules that allow microbes to communicate with each other and interact with our gut, and they are what constitutes a functional core.

In conjunction with core function, there are other hallmarks of a healthy gut microbiome. Though each microbiome is different and ever changing, the richness and diversity of the microorganisms living in the gut also contribute to proper function. *Richness*, in this context, refers to the total number of bacterial species present, while *diversity* is a measure of how evenly these species are distributed. In terms of insects, for example, richness refers to the number of different species present—flies, bees, butterflies, wasps, moths, fleas, and so on. However, if 90 percent of these insects are flies, this is not a diverse population, regardless of how many other species are represented. Researchers have found that high diversity of the gut microbiome is generally associated with health and stability over time.[7] Conversely, a lack of diversity makes the gut more vulnerable to infections and is apparent in many diseases, including obesity, inflammatory bowel disease, and type 1 and 2 diabetes. Not coincidentally, in the last several decades, we've lost critical microbial strains, and the diversity and richness of the gut microbiome in developed countries has been steadily declining.[8]

Diversity is also the main determinant of two other hallmarks of gut-microbiome health: resistance and resilience. Any population of microbes, insects, or people must have a degree of resilience to external or internal changes. If a gut microbiome can resist perturbation by pathogens, antibiotics, or a transient change to an unhealthy diet and if it is resilient enough to quickly return to its normal state afterward, it is generally considered healthy. Even if a gut microbiome provides all of the necessary core functions, without diversity and resilience, it's at a higher risk of disruption when faced with challenges.

The current understanding of microbial health is that it's not a fixed state but rather a dynamic and purposeful equilibrium. If dynamism is the key and everyone's microbiome is ever changing, how do we accurately identify unhealthy ones? Are there different definitions of microbiome health, depending where one lives? Researchers have sought to answer this question by charting variations in microbiomes in different parts of the world and among people pursuing culturally diverse lifestyles. Their studies have shown that the microbiome not only differs greatly from person to person, but also varies by population, geography, and time zone.

## Day-to-Day Variations of the Gut Microbiome

Christoph Thaiss, an assistant professor of microbiology at the University of Pennsylvania's Perelman School of Medicine, was working as a young investigator in the laboratory of Eran Elinav at the prestigious Weizmann Institute of Science in Tel Aviv when he found that the microbial ecosystem in both humans and mice is not static throughout the day and night. Instead, he found that there's a twenty-four-hour cycle of variation in its composition and function,[9] as well as in the interaction between it and the body. These fluctuations are influenced by when and what we eat, as well as by the suprachiasmatic nucleus (SCN), a tiny brain region in the front of the hypothalamus that functions as a clock or pacemaker, driving our circadian rhythm, the daily cycle of bodily processes that includes the sleep-wake cycle.

The circadian rhythm results from complex interactions among inputs and outputs to and from the SCN. The activity of nerve cells in this region fluctuates from day to night, varying the activity of the neurons and hormones that regulate many different body functions, including those of the gut and its microbiome. In addition, as in most complex networks, communication between the SCN and the gut is circular and made up of multiple feedback loops. Messages go to the brain, then back to the gut and its microbiome, changing microbial function, and then feed back to the brain again. The liver is usually kept informed of all communications on this channel. Disruption of this microbiome rhythmicity, which can occur when people change their sleeping or eating patterns, can make people more susceptible to disease, particularly metabolic syndrome.

To investigate the role of the microbiome in modulating these oscillations, Thaiss administered antibiotics to mice, thus suppressing their gut microbiomes and abolishing their oscillations, and he found that disrupting the function of the microbiome in this way caused serious interference with the activation of certain microbial genes. This interference resulted in the production of microbial signaling molecules, which entered the bloodstream and affected the function of several organs, including the liver and the brain.

Thaiss and his team also looked at the role of eating in relation to the circadian rhythm and found that the timing of our meals also plays a critical role in shaping microbial ecology and gut health.[10] When a person follows a normal pattern of eating during the day and not at night, the researchers found daily fluctuations of around 15 percent of the relative distribution of different microbial groups and a much higher percentage of their collective abundance. They were also able to confirm disruptions like those observed in their animal studies when they studied humans whose circadian rhythms were disrupted by jet lag. When the circadian rhythms are thrown off, so too is the gut microbiome. This study demonstrated for the first time that the rhythmic interactions between the gut connectome and the gut microbiome are synchronized with the sleep/wake cycle and mealtimes. Many of us know the surreal feeling of jet lag and its disorienting effects on concentration and sleep. Certain professionals, such as nurses, doctors, and police officers, induce this state regularly by their work schedules, but few are aware of the serious health challenges associated with these interruptions. Chronic disruption of the normal circadian rhythm, with its consequences on the rhythm of the gut microbiome, and the resulting changes in the communication among the gut microbes, the gut connectome, and our organs, is an important factor in the development of obesity, metabolic syndrome, chronic liver disease, and cognitive impairment. However, as I'll show in chapter 7, it is possible to counter this effect with a time-restricted eating plan to reestablish a normal rhythm in the microbiome and our own metabolism.

## Seasonal Changes in the Microbiome

The gut microbiome undergoes rhythmic changes by day and night, but research by Justin and Erica Sonnenburg's lab at Stanford University showed that oscillations also occur at a larger time scale, in synchrony with the seasons. The Hadza, an indigenous tribe who live in the central Rift Valley of Tanzania in East Africa, are descendants of an aboriginal hunter-gatherer population.[11] As of 2015, there were between 1,200 and 1,300 Hadza remaining in the world. It's believed that, until this last century, they've occupied their current territory for thousands of years with relatively few changes to their traditional hunter-gatherer lifestyle. Now, given the pressures of colonial administrations, tourism, and encroaching cattle farmers, only around 300 Hadza survive exclusively by foraging, bringing home honey, tubers, baobab fruit, other fruit, and wild game, based on seasonal availability.

Like all populations that live a traditional hunter-gatherer lifestyle unaffected by industrialization, the Hadza have avoided chronic diseases common in Western countries, such as obesity and diabetes. It's always important to bear in mind, however, that in addition to diet, other lifestyle factors—such as greater physical activity and lack of exposure to unnatural chemicals—likely also contribute to the absence or low rates of these illnesses.

There are two distinct seasons in the Rift Valley, which dictate the Hadza diet: a wet season from November to April, and a dry season from May to October. Even though they consume fiber-rich tubers and a variety of plants year-round, berry foraging and honey consumption are more frequent during the wet season, while hunting is most successful during the dry season, leading to a higher intake of lean meat during this time.

To study the effects of these seasonal changes on the Hadza microbiome, the Sonnenburg group examined 350 stool samples collected over the course of a year. They found, as reported in *Science* magazine in 2017, that the dry season, the period of higher game meat consumption, was associated with an increase in the bacterial phylum Bacteroidetes in the gut. This class of microbes declined by some 70 percent during the wet season, when the Hadza eat a more vegetarian diet, to a state that's much like the gut microbiota of people living in industrial societies.[12] However, in stark contrast to the permanent changes seen in industrialized societies, including the US, the Hadza microbiomes return to their full diversity again during the dry season. The gut microbes that are permanently reduced or extinct in industrialized societies return to detectable levels in the Hadza when their diet shifts.

The seasonal changes in the relative abundances of gut microbial taxa were accompanied by corresponding variations in carbohydrate utilization capacity—the supply of enzymes necessary to digest complex carbohydrates from animals, plants, and mucin (the key component of the mucus layer) into absorbable metabolites. During the wet season, the Hadza had lower levels of these enzymes while their abundance and diversity increased during the dry season. Samuel Smits, lead researcher on this study, suggested that these shifts in the microbiota's capacity to process different kinds of carbohydrates reflects the seasonal dietary changes and that the composition and function of a healthy microbiome for the Hadza changes with the seasons and the food consumed. This study clearly demonstrated that a healthy microbiome changes to match a person's dietary habits.

## Eating to Maintain a Healthy Microbiome

Following up the Hadza study, several research groups have studied gut microbial composition and function in other remaining indigenous communities with traditional lifestyles in South America, Africa, Nepal, and the Arctic, and they have found major differences between them and people in industrialized Western societies. While they have found that several aspects of industrialization may contribute to these differences, diet emerged as the most consistent factor.

In 2010 a research group from the University of Florence led by Carlotta De Filippo was able to prune the list of beneficial bacterial strains in a healthy microbiome by comparing the relative abundances of fecal bacteria from children aged one through six in Florence with those of children from a rural African village in Burkina Faso.[13] Children from the African village are breastfed, on average, until age two. In addition, all of the food they eat is harvested, cultivated, and produced locally. Their diet is low in saturated fat and animal protein but rich in starch, fiber, and plant polysaccharides, complex carbohydrates composed of a number of bonded sugar molecules. Much like early human settlements at the dawn of agriculture, their ways of eating are predominantly vegetarian, consisting mainly of millet and sorghum, black-eyed peas, and fresh vegetables. The carbohydrate, fiber, and non-animal protein content is very high.

In contrast, the Italian children were breastfed up to one year and later ate a typical Western diet high in processed foods, animal protein, sugar, starch, and fat but low in fiber. This modern Italian diet is markedly different from the largely plant-based Mediterranean diet that Italians used to enjoy, and which has been promoted as one of the healthiest diets in the world. The fiber content was about half that of the African diet.

It's no surprise that the diversity and relative abundance of microbial taxa differed considerably between the Italian and African children. The African children showed a greater abundance of the microbial phylum Bacteroidetes—the same microbes that increase among the Hadza during their hunting season, when their microbiomes are more diverse—and a reduction of the phylum Firmicutes. These two phyla comprise 90 percent of all microbial phyla living in the human gut.

The African children also had a greater abundance of bacteria from the genera *Prevotella* and *Xylanibacter*, which are members of Phylum Bacteroidetes. These discrepancies are critical because different microbes contain different genes that allow them to easily process the food to which they are adapted. *Prevotella* species have a set of genes that encodes for enzymes to digest certain plant fibers into short-chain fatty acids such as butyric, acetic, and propionic acids, molecules with many beneficial effects, such as maintaining the integrity of the gut wall, optimizing immune function, and signaling satiety. These short-chain fatty acids are a crucial component required for a healthy gut.

The same beneficial complex carbohydrate–metabolizing microbial genes found in the children of Burkina Faso have also been found among remnants of hunter-gatherer populations living in Africa (such as the Hadza) and South America (the Yanomami). It's no surprise that these genes were completely absent in the Italian children.

Studies of most traditional diets are performed in populations living in the Southern Hemisphere, but a 2017 study by investigators from the University of Montreal led by Geneviève Dubois examined the gut microbiome of the Inuit, who live a semitraditional lifestyle in the northern Arctic, mostly in Nunavut, the Canadian territory that is perhaps the last and by far the largest self-governing enclave of remnant hunter-gatherers on Earth.[14] Unlike other indigenous societies, however, the Inuit diet is similar in macronutrient composition to our Western diet in its high percentage of fat. The Inuit originally consumed a diet rich in wild animal and fish meat including seal, caribou, birds, and fish eaten raw, frozen, cooked, or fermented. They ate several types of seasonal plants and berries, but three-quarters of their calories came from animal fat. In contrast, Canadians living in Montréal get approximately 35 percent of their calories from fat and 50 percent from carbohydrates.

However, as in many places, Western lifestyles are encroaching on indigenous ways of life, and the Inuit diet today is a mixture of traditional and processed, shop-bought foods. The traditional foods, mostly animal-based, are eaten in summer and early fall, when hunting and foraging is easier, whereas Western foods are most popular in October and November. Dubois compared the Inuit participants with people of European descent living in Montréal who consume a typical Western diet and found a nearly 20 percent variance in microbiome composition between the two groups. In addition, the microbial variation within each individual was higher in Nunavut than in Montréal, possibly due to a more variable diet.

A compelling report was published in 2018 from a study conducted by researchers at the University of Minnesota in collaboration with the Somali, Latino, and Hmong Partnership for Health and Wellness. It demonstrated that the gut microbiota of immigrants from Southeast Asia rapidly Westernized, within months of their arrival in the United States.[15] In the US, the immigrants ate food richer in sugar, fat, and protein, and they "began to lose their native microbes almost immediately," said Dan Knights, the senior author of the study, testifying to the rapid adaptability of the gut microbes to environmental changes. "The loss of diversity was quite pronounced. Just coming to the USA, just living in the USA, was associated with a loss of about

15 percent of microbiome diversity." The obesity rates among many of these immigrant populations also grew severalfold. However, because the changes in diet occurred more slowly than those of the microbiome, American food alone cannot explain this rapid shift. Knights suggested that differences in drinking water (devoid of most microbes found in natural, untreated drinking water) and the use of antibiotics may have contributed as well.

The researchers found that in six to nine months the genus *Bacteroides*, more prevalent in industrialized societies, began to displace the non-Western genus *Prevotella* (both are part of the Bacteroidetes phylum). Overall microbiome diversity decreased more and more the longer the immigrants stayed in the US, and their children experienced another 5 to 10 percent loss of diversity.

This phenomenon, in which shifts in the adult microbiomes are amplified over generations, has been confirmed in preclinical studies too. For example, four generations of mice fed a low-fiber diet showed reduced microbiota diversity compounded over generations. Furthermore, restoration of a high-fiber diet didn't restore subsequent generations' diversity, suggesting that species within the microbiota of these mice had gone extinct during the four-generation length of the experiment.

## Western Diet is a Long-Term Stress on the Microbiome

Clearly the gut microbiome is highly responsive to diet. And as shown by my comparisons between Western diets and those of various traditional populations, the trend toward diets high in fat and refined sugar but low in fiber has had a serious impact on the vitality and diversity of the microbiome.

This shift has grown along with accelerating industrialization over the last seventy-five years, not only with the availability of cheap food components such as processed fat and sugar, but also with the addition of a long list of nonfood chemicals, such as preservatives, pesticides, herbicides, additives, and emulsifiers. It has increasingly affected the systems of the gut, which in turn affect the entire body. I strongly believe that this widespread dietary change is one of the main sources of our wide array of increasingly common chronic illnesses.

For one thing, the gut microbial ecosystem has been exposed to this stress for decades, pushing it beyond its resilient abilities, making it more vulnerable to new viral epidemics, and threatening its longtime symbiosis with the gut. The modern diet has not only reduced the diversity of microbiome species, but recent research from the Sonnenburg Lab shows that it has also significantly changed the relative abundance of several of the major remaining species.[16] When compared to gut microbes from traditional societies, those from industrialized parts of the world show a reduction of species from three microbial families (the Prevotellaceae, Spirochaetaceae, and Succinivibrionaceae) and an increase of several others. The latter include the microbial species *Akkermansia muciniphila*. These particular microbes reside in the mucus layer of the large intestine and have the ability to degrade the sugar molecules that make up mucus in the gut lining. In the absence of fiber—which is drastically reduced in the Westernized diet (the Hadza eat *ten times* as much fiber as the average American)—these microbes feed instead on the mucus layer, making it thinner and less effective, thus eroding the barrier between gut microbes and the gut lining.

The gut microbiome evolved along with our transition from hunting and gathering to farming to an industrialized society, but the microbiome was adaptive even before the advent of industrial agriculture. As humans migrated to different habitats and ate a diet based on the seasonal and geographic availability of various foods, the ever-shifting microbiome accommodated these changes. Now, however, the microbiome plasticity that helped us adapt may be undermining our health. As the gut microbiome raced ahead to adapt to dramatic changes in its environment, it has become less and less compatible with the rest of human biology, especially the more static gut connectome. This mismatch has resulted in dysregulations of the gut-based immune system—its acute inappropriate activation in autoimmune disorders and allergies, as well as the low-grade chronic activation seen in metabolic syndrome and in certain brain disorders.

These disruptions cannot be blamed on diet alone. Modernization, including valuable advances in medicine, has brought with it an increased and often inappropriate use of antibiotics and antiseptics, a higher level of sanitation (reducing benign microbes in drinking water), reduced contact with the soil and farm animals, and a greater number of cesarean sections—all of which have also contributed to the current microbiome changes. In fact, it's now well established[17] that exposure to antibiotics, stress, and compromised nutrition during the first thousand days of life can transform the gut microbiome for the rest of life.[18]

We can't turn back the hands of time or modernization, but we *can* alter our diets. Changing the way we eat and avoiding unnecessary use of antibiotics, in addition to adjustments in lifestyle, such as decreasing chronic stress and increasing physical exercise, are powerful and effective ways to wrest back control from our overwhelming crisis of chronic illness.

## Can We Test for a Healthy Microbiome?

Almost every patient I see comes to me with the hope of determining something conclusive about the health of their microbiome. Not only that, but they also request advice about the best prebiotics and probiotics to precisely address any deficiencies. Alas, it's not that simple.

There are a growing number of companies claiming that they can provide patients with a "microbial fingerprint" like a DNA test, a readout of the types and abundance of the various microorganisms in their guts, as well as an assessment of microbial function, if not give a personal diagnosis and treatment recommendation.

Sarah, a fifty-three-year-old journalist, is a good example of a curious patient who, feeling frustrated by her stubborn symptoms, decided to take matters into her own hands and send in her stool sample for analysis. By the time she consulted with me, Sarah had been dealing with persistent problems of abdominal bloating and distension, as well as an inexplicable, frustrating, fifteen-pound weight gain. This confounded her because, although she'd been eating a diet that included red meat, sugary drinks, and starchy carbohydrates like pasta and rice, she swore that she hadn't really made any changes to the way she ate over the last several years. She'd also been suffering vague symptoms, such as "brain fog," as she put it, and lack of energy. She described a hypersensitivity to certain foods, including gluten, dairy products, and lentils, as well as to many medications she'd been prescribed over the years.

After researching her symptoms on the Internet, Sarah had become convinced that these health challenges were related to poor gut health and the condition of her microbiome. She desperately wanted to figure out what was wrong and how she might lose the excess weight. She'd seen two other gastroenterologists before coming to me. The first attributed her symptoms to small-intestine bacterial overgrowth (SIBO) and "leaky gut syndrome." This doctor prescribed a course of Xifaxan, a non-absorbable antibiotic frequently prescribed for bloating symptoms. Sarah briefly felt better while taking the antibiotics—her bloating subsided, and she had more energy—but both of these symptoms returned a few weeks after the course was finished. The second doctor suggested that Sarah try the low-FODMAP diet—a diet low in fermentable oligosaccharides, disaccharides, monosaccharides, and polyols—which has become popular for patients with irritable-bowel syndrome and bloating.[19] By removing fermentable short-chain carbohydrates, such as those found in vegetable fiber sources like beans and other legumes, this diet deprives the gut microbes of their primary food source, resulting in reduced gas production and a reduction of the gas-related distension of her hypersensitive gut. Sarah noticed that her bloating improved, but not enough to continue such a restrictive diet.

Sarah also showed me her two reports from diagnostic tests based on her fecal microbiome analysis, one from the American Gut Project and one from a company called Viome. It's not unusual these days for my patients to arrive at their consultations with these or other commercial analyses of their gut microbiota in hand, hoping that I might help them decipher the results and offer a diagnosis with customized treatments. But at this point, these investigations aren't the equivalent of a blood test for cholesterol or sugar. Our science isn't there yet, as I explained to Sarah, although there is interesting and useful information to glean from these reports.

First, I looked over her results from the American Gut Project, a crowdsourced global citizen science effort cofounded in 2012 by microbiome pioneers Rob Knight, PhD, and Jack Gilbert, PhD, from the University of California at San Diego. I've often suggested that patients of mine send a stool sample to the American Gut Project. A participant contributes $99 and receives a kit to collect the stool sample. Each also answers a survey that includes questions about general health status, disease history, lifestyle, and diet. In return, patients receive a short report with graphs detailing what major taxa of microbes are living in their guts and how their results broadly compare to those of about twelve thousand other individuals throughout the world, as well as how they compare with others, for example, of the same gender, of a similar age, or on a similar diet. (There's even a chart comparing the client's microbiome with the extremely healthy one of author Michael Pollan!) The report also shows the four most abundant taxa, as well as the most enriched microbes—those with the largest amount of their favorite food in the person's diet. To tease out the identities and relative abundances of the bacteria living in a participant's mixed sample, the American Gut Project uses the standard analysis technique testing for 16S rRNA, a genetic marker unique to the prokaryotes, single-cell organisms without a nucleus—the bacteria and archaea. I believe this is an inexpensive, highly reliable test of the diversity and the relative abundances of microorganisms in one's gut.

However, it's important to bear in mind that the project's goal isn't to provide actionable information for a patient but to gain better scientific understanding of human microbiomes—which types of bacteria live where, the numbers of each, and how they're influenced by diet, lifestyle, and disease. In other words, this is a research project and makes no claims that its results will be useful in explaining symptoms or specifying treatment. All of the data collected by the American Gut Project is publicly available, allowing researchers around the world to mine it for meaningful connections between a person's microbial makeup and factors such as diet, exercise, antibiotic use, and lifestyle.

As we read her report together, it was clear to me that Sarah had a fairly typical reading. The relative abundance of the main gut microbial phyla, Firmicutes and Bacteroidetes, was similar to the average results of all the subjects in the database. Having significantly more Firmicutes than Bacteroidetes was consistent with Sarah's high-fat, low-fiber standard American diet. More important, the four most abundant taxa identified were the genus *Bacteroides*, the family Ruminococcaceae, the genus *Faecalibacterium*, and the genus *Blautia*—all taxa found in a healthy gut.

What I suspected, though this wasn't reflected in her report, was that Sarah was in the midst of perimenopause, when the ovaries begin to gradually make less estrogen. This is often associated with weight gain and bloating. Unfortunately, there's no easy explanation or treatment for these bothersome symptoms. Recently, however, my research group obtained funding from the National Institutes of Health to study the role of the gut microbiome in relationship to the dramatic changes in female sex hormones that occur during menopause. The goal of this study is to develop more effective therapeutic interventions for just the type of difficulties Sarah was experiencing.

I mentioned this to Sarah and told her that there is, in fact, evidence that links the gut microbiome to the complicated regulation of estrogen levels in the body. Other research suggests that the gut microbiome is part of a circular communication system that includes a woman's sex hormones, the liver, the gut, and many other parts of the body.[20] A change within this system, including a reduction of estrogen levels or a change in the abundance of gut microbes able to metabolize estrogens secreted into the gut before they're reabsorbed, will alter this communication, possibly leading to symptoms such as Sarah had.

Though she found the science interesting, Sarah still wanted an actionable plan for alleviating her symptoms, so we looked at her second analysis report, this one from Seattle-based company Viome. Instead of simply looking at the relative proportion of different microorganisms in the gut, they analyze the actual gene expressions of microbes with an aim toward precise personalized food recommendations. According to Viome, their ultimate goal is to prevent and reverse chronic diseases by identifying and treating the root causes, thought to be imbalances, inflammation, and dysbiosis in the gut microbiome.

Viome focuses on microbial function, one of the ways we're now able to categorize a healthy microbiome. They use a cutting-edge analysis method called metatranscriptomics, measuring the RNA (ribonucleic acid) that all microbial and human genes express. This is an excellent way of assessing gut microbial function, for transcription from microbial genes to RNA is a necessary step in the production of the molecules that microbes use to communicate with each other and the body.

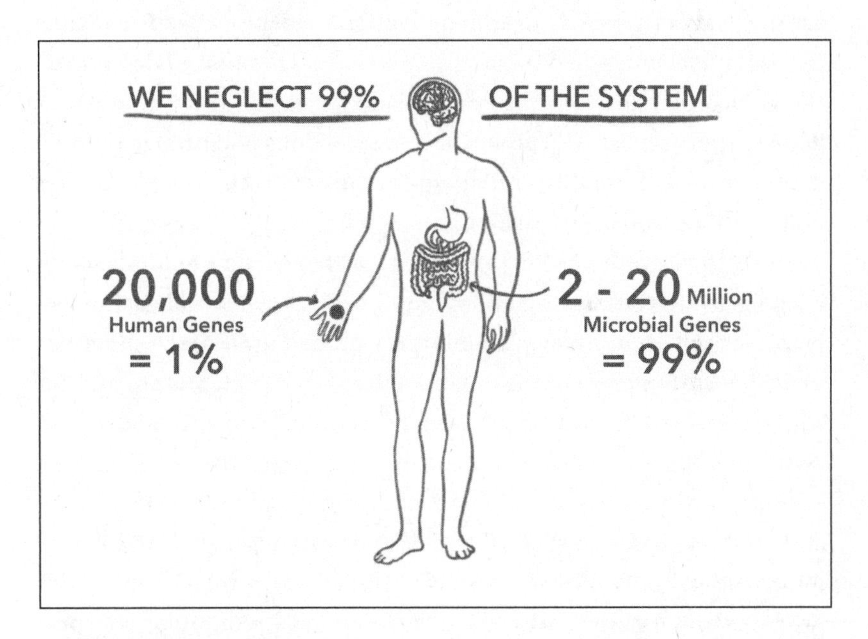


The Viome "Gut Intelligence Test" scores on a range of gut functions and labels these as either good or needing improvement. These gut functions are summarized under topics like digestive efficiency, intestinal-barrier health, overall gas production, protein fermentation, and metabolic fitness, among others. Sarah's "Gut Fitness Report" suggested that she needed improvement on several of these scores, including intestinal barrier health, inflammation, and overall gas production.

Viome combines analysis technology with artificial intelligence (AI) to produce personalized food recommendations. This section categorizes foods thus: superfoods, enjoy, minimize, and avoid. It also recommends specific dietary supplements, prebiotics, and probiotics. According to Viome, they've been able to train their AI engine using microbiome data from tens of thousands of people to make accurate predictions about which foods and nutrients are the best match for a particular patient's microbiome.

While I believe the methodology used by Viome to analyze microbiome function is state-of-the-art and superior to other comparable stool tests, its personalized diet and supplement recommendations are not currently based on publicly available science or well-designed clinical trials published in medical journals. Therefore, I don't think these recommendations are quite ready for prime time. However, I believe approaches like Viome's have tremendous potential to predict, diagnose, and treat chronic diseases in the near future and usher in a new age of personalized medicine.

In the meantime, I told Sarah that until sufficient published science backs up the recommendations, I prefer a more old-fashioned, empiric approach. Start with a largely plant-based diet, which has unequivocally been shown to correlate with health status, and for which a growing number of research studies have demonstrated significant health benefits in a variety of chronic diseases. The best example of this is the traditional Mediterranean diet. On such a diet, carefully watch out for foods that *consistently* increase digestive symptoms, and initially minimize or, if necessary, eliminate them. Dairy products and legumes, which can increase gas production, are commonly identified as symptom triggers—not surprising, considering that the majority of adults do not have the lactose-metabolizing enzyme lactase and that some legumes' metabolites can increase gas production in the large intestine. *Fermented* dairy products, such as yogurt or kefir, generally cause the least gas and bloating. I also recommended that Sarah cut down on all animal products, including dairy foods and red meat. A plant-based diet, providing large amounts of fiber from various plants, has been shown not only to increase the diversity and richness of microbes and the molecules they produce—including the health-promoting short-chain fatty acids—but also to offer an overall reduction in calorie intake because these foods have a lower caloric density; in other words, they contain fewer calories per given weight.

Let me illustrate this point with carbohydrates. If you consume one hundred grams of refined sugar, which doesn't contain any fiber and has a caloric density of four calories per gram, you will consume four hundred calories, and as all the sugar will be rapidly absorbed in the first part of the small intestine, none of these calories will reach the microbes in the large intestine and contribute to their diversity. On the other hand, if you consume one hundred grams of complex carbohydrates rich in fiber—like sweet potatoes, whole-wheat or oat bran, ancient grains, or wild rice, all of which have a lower caloric density—you will consume about seventy calories, or roughly one-sixth of the calories in the refined sugar, and the fiber that comes with such carbs will provide food for a variety of different microbes in your gut.

In addition to the Mediterranean diet, I recommended that Sarah add naturally fermented foods or drinks to her daily meals, begin a regular moderate exercise program, and practice time-restricted eating one day of the week (which I will discuss in more detail in chapter 7). I was able to convince Sarah that there was a fairly simple solution to her bloating symptoms and her weight gain. She had been relying on expensive tests and pre- and probiotic capsules.* I helped her to develop her own personalized diet to help her reduce her daily caloric intake and at the same time increase the diversity and richness of her gut microbiome.

---

* *Pre*biotics are nondigestible fiber and other food ingredients that promote the growth of beneficial microorganisms in the intestines. *Pro*biotics are the microorganisms themselves, which when administered in adequate amounts confer a health benefit on the host.

## Chapter Four

# STRESS AND BRAIN DISORDERS

Psychiatric and neurological disorders—such as depression, Parkinson's disease, and Alzheimer's disease—are among the most agonizing challenges afflicting us. Unlike other chronic illnesses, such as type 2 diabetes, obesity, metabolic syndrome, and cardiovascular disease, the rise of these disorders hasn't followed a straightforward trajectory. Diagnostic classifications of many psychiatric disorders have changed over time, so there aren't accurate measures of their prevalence over the last seventy-five years. Despite these limitations, research has pointed to a continuous increase of depression in younger populations and of Parkinson's, Alzheimer's, and autism spectrum disorder overall.[1] Nevertheless, there is a common factor in these metabolic, cognitive, psychiatric, and neurological illnesses: the gut microbiome.

Recent research has shown that individuals diagnosed with metabolic, cardiovascular, and mood disorders stand an increased risk of developing neurodegenerative brain disorders.[2] Chronic low-grade inflammation—stemming from an altered interaction between the gut microbiome and the gut-based immune system—is present in every one of these illnesses.[3] I believe the associated risk factors for brain disorders—neuroinflammation and vascular narrowing caused by chronic inflammation—are produced by the shift in our brain-body network that has taken place over the last seventy-five years. Just as various metabolism-related illnesses, such as heart and liver disease and certain forms of cancer, are manifestations of the disruptions wrought by industrialization, so too are brain disorders.

## Chronic Stress and the Brain-Gut-Microbiome Network

In my first book, *The Mind-Gut Connection*, I introduced the concept of a two-way conversation taking place along the brain-gut-microbiome axis. I've since adapted this theory to network science and refer to the brain-gut-microbiome network, or BGM network. This is an intrinsic part of the larger brain-body network (as mapped below). Communication in the BGM network is circular; information is sent in multiple feedback loops along two main trajectories—from the gut and its microbiome to the brain (bottom-up communication) and in the opposite direction, from the brain to the gut and microbiome (top-down communication). This bidirectional conversation profoundly affects the health of both gut and brain.

Just as modern lifestyles have facilitated a misalignment between

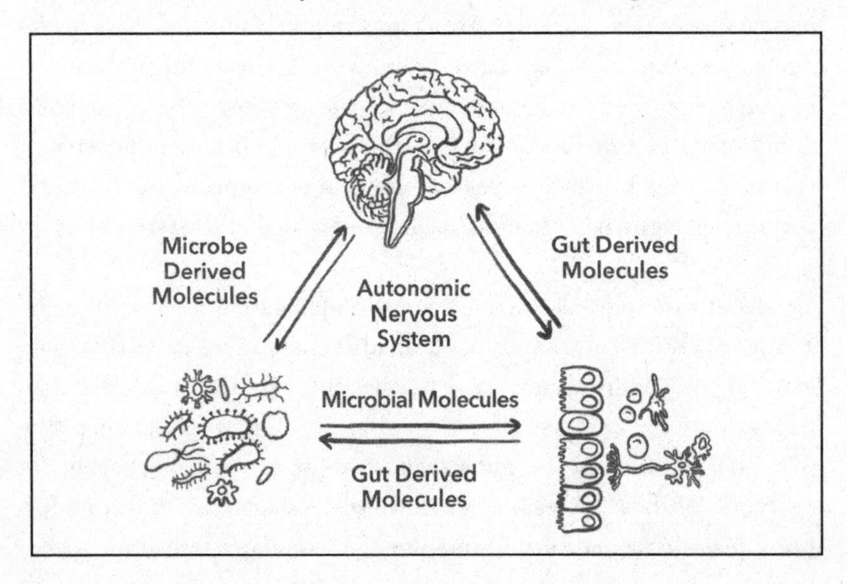

the gut connectome and the microbiome, so they have fostered a similar disparity at the brain level, producing another mismatch, this one between our ancient stress-response system and an unprecedented number of modern, generally non-life-threatening stresses. Occasional triggering of our nervous system's fight-or-flight response by immediate, life-threatening dangers once saved our ancestors from predators, and were essential for the survival of the human species, but today that response is triggered too often by less serious threats. This high level of perceived physiological stress and chronic anxiety has had drastic consequences, like disrupting the crucial communication lines of the brain-gut-microbiome network.

There are many well-documented effects of acute and chronic stress on this network, including a decrease in the abundance of *Lactobacillus*, an important microbial genus for maintaining gut health.[4]

Stress mediators like norepinephrine released into the gut can activate microbial genes that increase bacterial engagement with the gut immune system. Stress has also been shown to change contractions and peristalsis throughout the gut, influencing the time it takes for the contents to travel through different regions and in turn affecting the habitat and food supply of the microbes. Moreover, stress has been shown to increase intestinal permeability, creating what is popularly

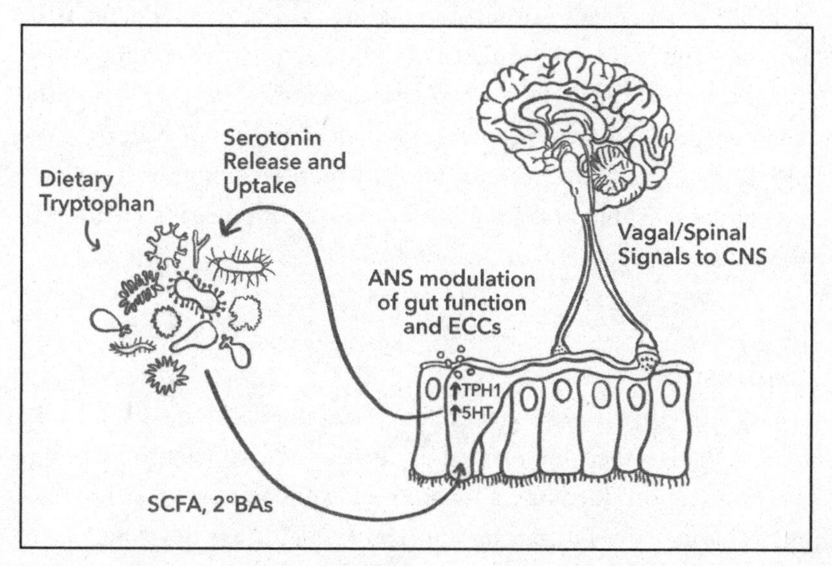

known as a "leaky gut," which can result in low-grade activation of the gut's immune system.[5]

It's clear that many changes in gut health that have in the past been attributed solely to an unhealthy diet are likely caused in part by chronic stress signals sent by the brain to the gut, altering gut microbial signaling molecules that then send alarming feedback to the brain. Complex brain disorders cannot be adequately explained by simple linear concepts, but rather require the more comprehensive lens of network science, which shows that chronic stress and anxiety combined with an unhealthy diet and lack of regular exercise exert a synergistic detrimental effect on gut health. This two-pronged influence on the BGM network has created a continuous, and continuously altered, feedback loop among the brain, the gut, and its microbiome.

And yet, despite the fact that we're all exposed to the same changes from industrialization, not all of us will develop a chronic brain disorder. Our vulnerability to neurological illness is influenced by genetic factors and by epigenetic programming early in life. These factors influence the structure of each individual's BGM network, which determines its vulnerability to disturbances throughout life.

Though largely based on animal models, microbiome alterations have been implicated in the development of almost every brain disorder in the last decade—from anorexia to schizophrenia. I'm focusing on the recent research for three of them here: depression, neurodegenerative disorders (Parkinson's disease and Alzheimer's disease), and neurodevelopment problems (autism spectrum disorder). These are illnesses that have not only steadily increased in prevalence over recent decades, but also have been definitively linked to changes in the gut microbiome and to chronic stress.

## Depression

About fifteen years ago, scientists, researchers, the media, and the general public became excited about the new concept of a mind-gut connection, focusing almost exclusively on how an altered gut microbiome—the bottom-up contribution—fosters development of several brain diseases. This was, in large part, based on paradigm-shifting results from animal studies, such as the ones showing that "germ-free" mice, entirely without gut microbiomes, exhibit abnormal emotion-like behaviors and learning and memory deficits.[6] These eye-opening studies established that the signals the brain receives from the gut and its microbiome can modulate brain function and behavior. Researchers ramped up efforts to determine if changes in the number, diversity, or function of gut microbes could be linked in some way to major depressive disorder.

In recent years, these studies have been followed by a series of striking experiments in which fecal matter from depressed people was transferred into germ-free mice or antibiotic-treated rats. The animals began to show signs of despondency that are generally considered to mirror aspects of human depression.[7] Such experiments have brought us a giant step forward, unequivocally demonstrating that the fecal transfer of microbes and their metabolites from humans can change mouse behavior and brain biochemistry, but we still haven't determined if any of these metabolites, on their own or in combination with other molecules such as inflammatory signals, are actually causing depression in human subjects.

Nonetheless, many researchers continue to seek a universal "microbial depression signature," in which specific metabolites produced by the gut microbiota could be linked to depression in humans, as opposed to approaching this ailment as a biological system disturbance in which the BGM network plays a prominent role. Fortunately, there *have* been a few recent studies focused solely on the bottom-up loop, which have made great strides in helping us better understand depression.

In 2015 a research group led by Hai-yin Jiang, from the State Key Laboratory for Diagnosis and Treatment of Infectious Diseases of Zhejiang University in Hangzhou, China, established that research subjects could be classified as depressed or depression-free solely by their gut microbial composition.[8] More specifically, when researchers compared the microbes in stool samples from forty-six patients with a diagnosis of major depressive disorder and thirty healthy control subjects without depression, they found that the depression group had a greater percentage of the phyla Bacteroidetes, Proteobacteria, and Actinobacteria but reduced levels of *Faecalibacterium*, a genus in the phylum Firmicutes. A greater abundance of Firmicutes, as was found in the Hadza microbiome, is generally thought to be a marker of a healthy gut and is also associated with anti-inflammatory properties. Researchers found that the more of these beneficial microbes were present in the stool, the less depressed the patients were.

In 2016, two other experiments took this finding one step further. One was done by researchers collaborating from various scientific institutes in Chongqing, China; the other, by researchers at the APC Microbiome Ireland at the University College Cork. The researchers transferred fecal matter from people with major depression into germ-free mice (in the Chinese study) and into antibiotic-treated rats (in the Cork study); in other words, none of the animals had intact microbiomes. The researchers were aiming to demonstrate that the altered gut microbiome identified in depressed patients was directly responsible for their downward shift in mood.[9] In both of these experiments, the animals that received fecal transplants behaved in a depressionlike manner; in other words, their behaviors mirrored traits exhibited in human depression. In addition, the "depressed" mice showed disturbances in microbial genes like those observed in depressed people. Both studies suggested that certain microbial metabolites from the guts of depressed patients, such as the tryptophan metabolite kynurenine, influenced demeanor in the mice, producing the behaviors and brain function of anxiety.

Yet all three of these studies came up with varying, at times contradictory, comparisons between the gut microbial composition of depressed patients and that of healthy patients studied as controls. One, for example, reported a *reduction* in Bacteroidetes in depressed patients, the opposite of the finding reported earlier by Jiang. In other words, none of these researchers were able to locate the elusive universal microbial depression signature that would prove a direct link between specific gut metabolites and depression.

I believe there's a limit to how much information we can translate from studies of laboratory mice. The difference between mice and humans is too vast. Lab mice are inbred, rendering them genetically indistinguishable. They're all raised under identical conditions, eating the same food, living at the same temperature, experiencing the same early life circumstances. Moreover, the few test strains used for the fecal transplants are selected from hundreds of different strains, all of which vary in the biological mechanisms, gut microbial composition, signaling molecules and neuroactive metabolites, and receptors playing a role in the brain-gut-microbiome network. And importantly, the complexity of the human brain and its role in generating emotions is vastly different from the mouse brain.

The humans enrolled in these studies, on the other hand, differ from one another in *every* way—genetically, environmentally, and microbially, not to mention that they have different diets and a wide range of life experiences. The observations obtained in a small group of germ-free mice need to be assessed and confirmed in tens of thousands of patients before researchers could reasonably draw definitive conclusions. Fortunately, however, such large-scale studies for several brain disorders are now under way, and I'll discuss them later in this chapter.

It's no small feat to demonstrate that some of the signaling molecules generated by the gut microbiota as they metabolize food components, such as amino acids and molecules secreted into the gut, like bile acids and hormones, may be linked to depression. However, results obtained in these studies will be meaningful only when we're able to fully understand *how* such compounds affect the function of the BGM network. We do already know that thousands of such molecules produced by the gut microbes directly as well as many inflammatory mediators originating from interactions of the microbes with the gut-associated immune system are participating in the communication within this complex network. Still, we are a long way away from understanding the precise code of this communication, and even further away from identifying therapeutic targets.

Recent research on the neurotransmitter serotonin *has* taken such a systems approach and made extraordinary findings. Most serotonin is made in the gut with the aid of the microbiome, but a small amount can also be produced by the brain independently. Serotonin is well known as a molecule that plays a significant role in sleep regulation, pain sensitivity, appetite, and other vital functions; it has also been implicated in several brain disorders, in particular depression and ASD. In the gut, serotonin helps regulate intestinal motility and secretion. Made from the essential amino acid tryptophan, gut serotonin, along with two other neuroactive tryptophan metabolites, kynurenine and indole, are some of the most extensively studied signaling molecules in the BGM network.[10]

Despite the critical role serotonin plays in fine-tuning the brain's functions, less than 5 percent of it is produced and stored in the brain. This small amount is found in brainstem nerve cells that send ascending projections to practically all regions of the brain as well as descending projections to the spinal cord; therefore due to these extensive projections it has a vast, wide-ranging influence on neurological activity and behavior. Its influence is dominant in our emotion-regulating networks, and so helps modulate our moods. This is the premise of selective serotonin reuptake inhibitors (SSRIs), a class of antidepressant drugs that are widely seen as the most effective pharmaceutical intervention for depression. SSRIs were created to increase the concentration of serotonin in various brain regions.

The other 95 percent of the serotonin found in our bodies is created and stored in special cells contained in the lining of the gut called enterochromaffin cells (ECCs), which function as serotonin warehouses, as well as in a small number of nerve cells in the enteric nervous system. The ECCs are distributed throughout the small intestine and colon.

When stimulated by microbes or by intestinal contents moving through the gut, the ECCs secrete serotonin in the gut wall, both onto sensory nerve endings and into the circulation, as well as into the gut lumen. However, while released serotonin has powerful local effects on the gut and indirectly on the brain, there is very little serotonin reaching the bloodstream, as it is rapidly taken back up by the ECCs and by platelets in the blood. In addition, serotonin is unable to pass the blood-brain barrier—a layer of cells that keeps most molecules in the circulation out of the brain. And yet, serotonin released in the gut can have important influences on brain function, because important targets of gut-secreted serotonin are the vagus nerve's sensory endings. When stimulated, they generate long-distance vagal signals to emotion-regulating networks in the brain,[11] so serotonin *is* able to signal to the brain in this way.

Despite the fact that brain- and gut-produced serotonin have long been considered distinct from each other, recent studies have shown that our gut microbes, in response to the food we eat, can influence the synthesis and secretion of serotonin in the gut, so these microbial actions may have important implications for the brain and many of our vital functions, such as pain sensitivity, sleep, and appetite.[12] The communication between the food we eat, the gut microbes, and the gut is a two-way street. The microbes provide an important stimulus for serotonin production in enterochromaffin cells, and a portion of this serotonin is secreted into the inside of the gut, the lumen, where it can affect the microbes. Recent research suggests that this luminal serotonin plays an important role as mediator between the microbiome and the gut cells.

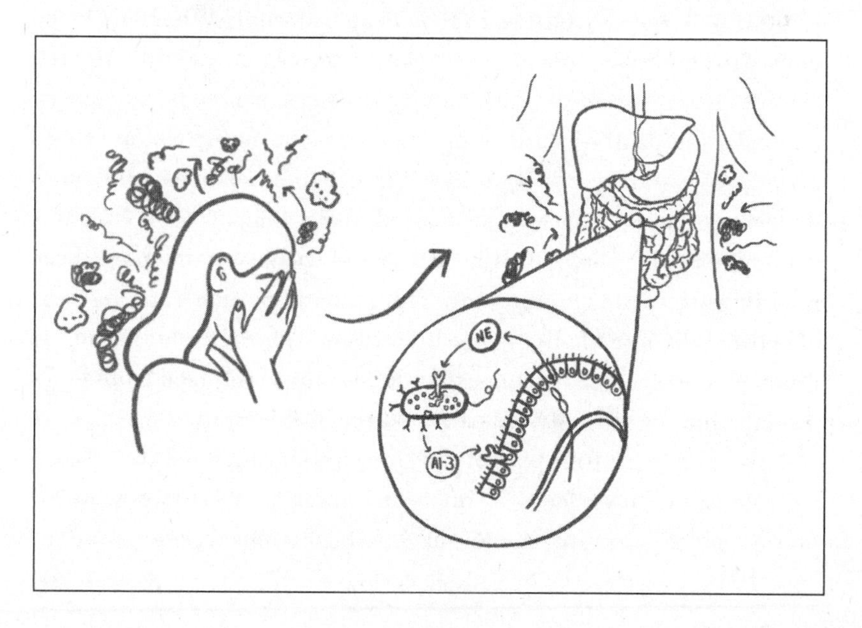

One approach that investigators are using to better understand the role of microbes in the regulation of serotonin and other tryptophan metabolites is to compare study results between mice raised under sterile conditions, or germ-free mice, and normal laboratory mice. In one such study, scientists found that the amount of serotonin in the circulation of germ-free mice was only *half* the amount found in mice with a normal microbiome. Furthermore, the higher serotonin concentrations in normal mice were accompanied by enriched expression of genes necessary for serotonin production. These findings are evidence that certain gut microbes regulate serotonin synthesis and serotonin signaling throughout the brain-gut network.

My colleague and collaborator Elaine Hsiao, PhD, assistant professor in the Department of Integrative Biology and Physiology at UCLA, has demonstrated in a series of elegant experiments that short-chain fatty acids (the microbial breakdown products of dietary fiber) and secondary bile acids (the microbial metabolites of the molecules that facilitate fat absorption) are responsible for nearly half of serotonin synthesis in ECCs. This unexpected feat is accomplished through the microbial stimulation of a particular enzyme located within ECCs, which enables the first step in metabolizing tryptophan from food into serotonin. Based on findings in Hsiao's mouse studies, the more tryptophan the microbes encounter—in chocolate, oats, dates, milk, yogurt, cottage cheese, red meat, eggs, fish, poultry, sesame seeds, chickpeas, almonds, and sunflower and pumpkin seeds—the more these microbes provoke serotonin production in the ECCs. In other words, the more we feed our gut microbes plant-derived fiber's complex carbohydrates and tryptophan-rich foods like cheese and chocolate, the more these microbes stimulate gut-serotonin output, with its wide-ranging benefits throughout the body. Still, while this peripheral system is highly productive, there's no evidence that the serotonin released from ECCs directly into the bloodstream can cross the blood-brain barrier.

This is only the first part of the story of the relationship between what we eat, what we feed our microbes, and the role they play in the production of serotonin in our gut. Dr. Hsiao's most recent study revealed a captivating plot twist: microbes themselves are also affected by the very serotonin that they stimulate. Hsiao found that certain microbes have a molecule in the cell membrane that is very similar to the molecule in the cell membrane of blood platelets and brain cells that enables these cells to take up serotonin. This serotonin transporter is the same molecule expressed by nerve cells in the brain that is the target for SSRI antidepressants such as Celexa and Prozac.[13]

In other words, serotonin released by ECCs into the lumen of the gut can be taken up by microbes, altering their behavior. Early rat studies had already shown that acute stress can release serotonin into the gut lumen,[14] but scientists, prior to the advent of microbiome science, had a difficult time coming up with a reason for this finding. However, we now know the reason why nature has come up with this communication between the ECCs, the gut lumen, and the microbes. Microbes cannot produce serotonin from tryptophan themselves. Only the ECCs and cells in the brain are able to make this conversion, so the lumen is the only source of serotonin for them. Though the consequence of this microbial uptake of serotonin for our health remains unknown, it has led to the intriguing speculation that antidepressants exert their effect not only on the brain, but also, by increasing the amount of serotonin in the gut lumen, on the communication between ECCs and microbes. This increase in serotonin levels in the gut may play a role not only in the common gastrointestinal side effects reported by patients taking SSRIs, but also in some of the unique clinical features of antidepressant therapy. For example, differences in the gut microbiome, as well as in diet, may underlie the differences in individual response and side effects to this group of medications and could explain the slow onset of therapeutic effects and the continuation of effects long after the last dose. In addition, the involvement of gut microbes in serotonin physiology could explain the beneficial effect of dietary interventions, as a supplement to medication, in depressed patients like Sarah.

Science suggests not only that gut microbiota are key players in stimulating serotonin production in the gut ECCs but also that their involvement in the breakdown of tryptophan into neuroactive molecules is much more widespread. One example of such a compound with direct relevance for brain function and brain diseases is the tryptophan metabolite kynurenine. In the GI tract, kynurenine is synthesized from tryptophan by the enzyme indoleamine-2,3-dioxygenase (IDO).[15] This gut-based enzyme and the associated production of kynurenine are greatly influenced by the health of the gut and by the activity of certain microbes. One class of microbes from the phylum Firmicutes plays a crucial role in regulating the synthesis of serotonin in the gut ECCs, but a different species (*Lactobacillus*, another member of the Firmicutes) determines how much of the tryptophan is converted to kynurenine.

While most readers are familiar with the amino acid tryptophan and the "happiness molecule" serotonin, few will have heard of kynurenine, yet it plays an equally important but opposite role in the effects of chronic stress on our body and brain. A large number of scientific publications have implicated its dysregulation in several brain diseases, including depression and Alzheimer's disease. For example, chronic stress in mice, rats, primates, and humans has been shown to lessen the relative abundance of *Lactobacillus*.[16] In rats, it has been shown that this decrease reduces the animal's ability to break down tryptophan into serotonin. Chronic stress is accompanied by an increase in the enzyme IDO, which leads to an increase in kynurenine, which, unlike serotonin, can freely cross from the blood into the brain. Some of the most consequential effects of increased kynurenine in the brain are neuroinflammation and neurodegeneration, both implicated in some forms of depression and in Alzheimer's disease.[17] Moreover, because kynurenine competes with tryptophan to cross the blood-brain barrier, the more kynurenine produced in the gut, the less tryptophan is available in the brain to make serotonin. An increase in the ratio of kynurenine to serotonin has been implicated in Alzheimer's and some forms of depression. The upshot is that decreasing chronic stress and making dietary changes that induce gut microbial abundance and function may reduce the amount of tryptophan that is turned into kynurenine, shifting the balance toward serotonin synthesis. Studies are currently under way to determine if such a shift may have a therapeutic effect in several brain disorders.

Although the metabolism of dietary tryptophan into serotonin and kynurenine is accomplished by cells in the gut and modulated by microbes, *only* the gut microbes themselves are able to metabolize tryptophan into another group of metabolites, the indoles. Indoles are a large number of closely related compounds, which have a wide range of functions in the human body and brain. For instance, Vadim Osadchiy, a student researcher in my lab, has recently demonstrated that one indole metabolite may help regulate a brain network that influences our desire to eat.[18] Another one, indoxyl sulfate, has recently received attention, as it may play a role in the development of autism spectrum disorder, Alzheimer's disease, and depression.[19]

The recent discovery of the involvement of several of these tryptophan metabolites in different brain and gut disorders reinforces the concept that molecules derived from tryptophan and modulated or generated by the gut microbes play crucial roles in our brain-gut-microbiome network, and that changes in this complex communication system can develop as the result of an unhealthy diet, chronic stress, or both.

## Neurodegenerative Diseases

Alzheimer's and Parkinson's diseases are the two most prevalent neurodegenerative diseases in the world. Alzheimer's is characterized primarily by memory loss, while Parkinson's symptoms, such as tremors and slow movement, are motor related. Although the manifestations are different, both illnesses are characterized by the presence of certain abnormal proteins in the brain—beta-amyloid plaques and tau-neurofibrillary tangles in Alzheimer's and alpha-synuclein Lewy bodies in Parkinson's. Both diseases, however, share multiple symptoms, such as depression, anxiety, sleep abnormalities, and cognitive impairment. One possible explanation for this overlap involves a tiny structure in the brainstem, the locus coeruleus (LC, literally "blue-spot"), which produces the hormone and neurotransmitter norepinephrine, important for the regulation of attention, arousal, and mood. Researchers have suggested that degenerative changes in the LC might play a role in the neuropsychiatric abnormalities shared by those suffering from Alzheimer's and Parkinson's.[20]

There's another structure in the same region of the brain, with close connections to the LC, known as the nucleus tractus solitarius (NTS), the "lonely trail kernel." This structure receives sensory signals from the vagus nerve, a major communication pathway between the gut, its microbes, and the brain. In Parkinson's patients, the NTS has been shown to exhibit neurodegenerative changes.[21]

These tiny interconnected brainstem structures, the LC and NTS, are crucial relay stations along the BGM network, consistently delivering messages between the microbes in the gut and the brain. Changes in the information flow through these structures are likely to be involved in both Alzheimer's and Parkinson's diseases. The BGM network and its many stations along the way, crucially implicated in the development of these neurodegenerative illnesses, are currently being explored intensively by investigators around the world, offering striking new insights into the connection between our diet and the health of the gut, the brain, and the mind.

## Parkinson's Disease

Parkinson's disease has generally been considered a strictly neurological disorder, its fundamental difficulties manifesting in movement and gait, but Parkinson's patients also suffer from a wide range of nonmotor, gut-related symptoms, including constipation and indigestion. These symptoms, such as the slow passage of waste through the large intestine, delayed emptying of food by the stomach, and increased sensitivity to gut stimuli, are often attributed to dysfunctions in the autonomic and enteric nervous systems.[22] The risk of developing Parkinson's has been found to increase as the number of bowel movements declines and the severity of constipation increases. In close to 40 percent of patients, these are among the earliest signs of the disease, appearing as early as fifteen years before any clinically detectable neurological and motor-related symptoms.[23]

There is now, in fact, a growing body of fascinating research suggesting that certain gut-microbiome symptoms may precede the neurodegenerative changes of Parkinson's disease in the brain by more than a decade.[24] Although clinical studies remain limited to observing differences between patients and healthy subjects, it's plausible that, when they occur in otherwise healthy people, the large variety of gut microbial changes researchers have found among Parkinson's sufferers increase the risk of developing the disease. For example, a number of studies have confirmed a reduction in Parkinson's patients of the abundance of Prevotellaceae, the family of microbes containing the *Prevotella* species.[25] As discussed in chapter 3, this shift has also been observed in industrialized societies consuming a Western diet.

Among the changes caused by the reduction of *Prevotella* and the associated decrease in short-chain fatty acids (SCFAs) is a reduction of intestinal mucus production, compromising the gut's barrier function. Along with the *Prevotella* decline, Parkinson's patients have a decreased production of SCFAs from dietary fiber, important for gut health including the integrity of the intestinal barrier.[26]

Such observations have so far been limited to correlations between diet, microbiota, and disease but have not proven that changes in the microbiome are actually one of the causes of the disease. Despite the popular media's tendency to present this research as a breakthrough, the studies are, for now, inconclusive.

Nevertheless, researchers have found that, like mice with depression-like symptoms, mice with Parkinson-like behaviors become more impaired when given fecal transplants from humans suffering from Parkinson's disease, but they do *not* after fecal transplants from healthy humans. Other research shows that gut microbes can have several negative influences on the brain-gut-microbiome network, including effects on immune and nerve cells in the brain, as well as on the integrity of the blood-brain barrier and on intestinal permeability.[27] Considering these findings together, one can speculate that the early gastrointestinal symptoms reported by some patients with Parkinson's may actually be the first indicators of changes in the interactions between the gut microbiome and the gut connectome. And with the brain.

Researchers have recently found that the gut microbiome may play a role in treatment of Parkinson's disease as well. A 2019 study by a group of scientists at the University of California in San Francisco and at Harvard University, under the leadership of Dr. Emily Balskus from the Department of Chemistry and Biological Chemistry at Harvard Medical School, demonstrated that the primary medications used to treat Parkinson's disease affect patients variously depending on the composition of their gut microbiota.[28] The motor symptoms of Parkinson's—muscle stiffness, altered posture, disturbed gait, involuntary movements, and tremors—occur when there's a deficiency of the neurotransmitter dopamine in a particular region of the brain. The primary medication for this disorder, levodopa, or L-dopa, enters the brain, where it's metabolized into dopamine by a specific enzyme. (L-dopa can cross the protective blood-brain barrier, whereas dopamine cannot.) However, only 1 to 5 percent of L-dopa actually ever reaches the brain, because it's first metabolized at different sites within the body, in particular by certain strains of microbes living in the gut. To allow more dopamine to enter the brain, doctors generally prescribe a second medication, carbidopa, which helps to block L-dopa from being metabolized in the gut before it can enter the brain. Unfortunately, in patients with Parkinson's disease, carbidopa was found to be largely ineffective in preventing the gut microbiota from metabolizing L-dopa. Even with this dual medication strategy, nearly 60 percent of L-dopa is inactivated by the gut microbes.

In their investigation of this phenomenon, Dr. Balskus and her team identified certain strains of a microbial species, *Enterococcus faecalis*, that play a crucial role in the metabolism of L-dopa. Depending on its abundance, genetic makeup, and the enzymes it produces, the efficiency of the breakdown of L-dopa into dopamine in the gut can vary widely. Because everyone has a different microbiome, with as little as 10 percent of microbial strains shared from one person to another, the response of Parkinson's patients to these treatments can vary significantly.

Given the emergent evidence that microbes modulate our responses to many drugs, from SSRIs to L-dopa, and that diet influences our gut microbial composition, dietary interventions may be useful in some patients with Parkinson's disease and other brain disorders. In L-dopa therapy, sometimes a specific microbial strain can be targeted to create an environment in which less of the drug is broken down in the gut. In patients with depression, there might be beneficial effects to supplementing SSRIs with a Mediterranean diet, perhaps improving the interaction of the medication with the microbial mechanism.

## Is It Possible to Slow the Progression of Neurological Diseases?

Ever since I told the story of a patient with early-onset Parkinson's in my book *The Mind-Gut Connection*, an increasing number of patients have come to my office seeking answers about the role of the gut and its microbiome in this neurological disorder. They primarily want to know if anything can be done to slow the progression of this insidious disease. David, a fifty-five-year-old farmer from Fresno, California, was one such patient, albeit an unsuspecting one, as he didn't yet know his diagnosis when we first met. Thankfully, we were able to identify his illness long before its neurological manifestations had fully developed. Initially, David came to my clinic with his wife, Cindy, to discuss his health more generally. As we talked, it was clear to me that this couple had weathered the inevitable storms of a long marriage well. They sat close together, chatting amiably about their lives on the farm, their three children, and their recent concerns about David's health.

David spoke first about his medical history. Except for some weight gain and an increase in his blood pressure and cholesterol levels five years ago, for which he was now taking medication and a statin drug prescribed by his primary-care doctor, he'd been generally healthy throughout his life. I asked David and Cindy about their lifestyles.

Though the family was fairly active, given the demands of the farm, they'd always eaten a typical Western diet, based on sugary breakfast cereals, bacon and eggs, and regular consumption of red meat, potatoes, and bread. Salads and other vegetables, they admitted, made only rare appearances at their dining table.

David added that, even though he'd never had problems with his bowel movements in the past, he'd recently noticed that he no longer kept to his daily routine. Sometimes he would skip one or even two days without defecating. "I'm not too worried about it," David said, then pointed to his wife and smiled. "She was the one who wanted me to see a specialist." After he'd had a normal screening colonoscopy, he said, his primary-care doctor assured him not to be concerned and prescribed a laxative he was meant to take only if his symptoms worsened.

"There *is* one more thing," Cindy added when David had finished speaking. "Over the last couple of years, there have been times when I've woken up in the middle of the night because David is talking loudly in his sleep, sometimes even yelling. One night, I saw him actually jump out of bed and begin walking around."

He often seems distraught then, she explained, as if he's having a terrible nightmare, which he confirmed. "Anyways, I thought it was worth mentioning since it's so unusual and David never did this before."

It was uncanny to me that David and Cindy had brought up these seemingly unrelated symptoms, as if they'd somehow known that these two details fitted together to create a complete picture of David's health. Both of these symptoms, as I'd recently seen in a couple of other patients, can be early harbingers of Parkinson's disease. A new onset of constipation and REM-sleep behavior disorder have been identified in patients some ten to fifteen years before the typical neurological symptoms.

Normal sleep has two distinct states. The first is slow-wave sleep, the lighter phase that precedes the deeper stage of rapid eye movement (REM).[29] This second phase is when dreaming occurs and the brain is highly active; in fact, the electrical brain activity recorded during REM sleep is similar to that recorded during wakefulness.

Most people think dreaming is a purely mental activity, but dreamers are also experiencing temporary muscle paralysis during this time to prevent body movements associated with the dream, which might awaken them. In persons with REM-sleep behavior disorder, however, this paralysis is incomplete or absent, allowing them to act out their dreams without waking up. Some people can even engage in normal daytime activities. While this sleep abnormality is relatively rare, in a University of Minnesota Medical School study published in the journal *Neurology*, 38 percent of patients diagnosed with it developed Parkinson's disease in an average of twelve to thirteen years.[30] As it was with David, when both this sleep abnormality and a new onset of constipation were present at once, the likelihood doubled.

These were not the only clues from David and Cindy that guided me toward a diagnosis of Parkinson's. During our conversation, Cindy mentioned that their farm was located in California's Central Valley, once often called the richest agricultural region in the history of the world. It's a four-hundred-mile-long swath of some of the world's most productive farmland. It's also the epicenter of industrial agriculture, as about one-fourth of the produce consumed in the United States is grown there—and nearly half of all pesticides, herbicides, and fungicides used in this country are sprayed on crops in this region as well. In fact, during our meeting, Cindy described having to regularly and urgently call her children into their house when she heard the distant aerial buzz of crop dusters approaching, to shield them from the chemical rain.

Not coincidentally, this also happens to be one of the areas in California with the highest prevalence of Parkinson's. My UCLA colleague, Dr. Beate Ritz, professor and vice chair of the Epidemiology Department at the School of Public Health at UCLA, with co-appointments in the Environmental Health and Neurology departments, actually conducted a study about Parkinson's disease in the county where David and Cindy live. Dr. Ritz and her research team enrolled 368 patients diagnosed with Parkinson's disease between 1998 and 2007 who had lived in California's Central Valley for at least five years prior to diagnosis, and they gathered an equal number of healthy control subjects.

They then collected estimates of residential exposure to two common pesticides (maneb and paraquat) between 1974 and 1999. They found that those who'd been exposed to both pesticides within five hundred meters (about one-third of a mile) of their home had a 75 percent increase in risk for developing Parkinson's. This risk rose more than fourfold for people under sixty at the time of diagnosis, meaning they would've been children, teenagers, or young adults during their period of exposure.[31]

There've been other epidemiologic studies demonstrating that pesticides increase the risk of Parkinson's. In fact, over the last two decades, evidence has shown that pesticides produce some of the neurochemical, behavioral, and pathological features of this disease in animals as well. Pesticides and herbicides are engineered to be toxic to keep pests and weeds at bay; neurotoxins, for example, paralyze insects. Many others inadvertently contribute to loss of nerve cells by compromising various biologic systems, especially in the gut. This was recently shown in studies done in laboratory mice looking at the effect of the insecticide diazinon, used to control insects on fruit, vegetable, nut, and field crops.[32]

I told David and Cindy about my presumptive diagnosis, explaining that the possibility of long-term exposure to the chemicals used on their farm may have triggered early signs of Parkinson's in David's gut and brain. I referred him to a UCLA neurologist with expertise in this disease to confirm my initial diagnosis. I also explained that even though there are currently no effective medications to slow the progression, the fact that David may have discovered this up to fifteen years before developing the full-blown neurological manifestations increased his chances that treatments targeted at the gut microbiome will have been developed in the meantime.

I wasn't simply being optimistic: several biotech companies are currently working on new treatments for Parkinson's aimed at the gut microbiome, and my research group is also working with Dr. Ritz on a project funded by the National Institutes of Health to explore the role of the gut microbiome in the development of the disease. I'm genuinely hopeful that progress will be made in the near future.

I also recommended to David—despite the fact that this is not yet supported by conclusive scientific evidence—to shift to a largely plant-based diet high in fiber, polyphenols, and omega-3 fatty acids, all of which have demonstrated beneficial effects on the gut and the brain. Specifically, I explained that the intake of dietary fiber may increase the relative abundance of the *Prevotella* microbes, which in turn may increase the availability of SCFAs in the gut. For patients in the throes of this type of illness, a variety of approaches—diet, behavioral therapy in addition to medication—are necessary to address all aspects of the body's network system.

Despite the worrisome possibility I was raising, David and Cindy were both gracious, taking in the news carefully and promising to follow up with the specialist I'd recommended. I didn't see them again, but Cindy called me several years later to tell me that David had indeed received a diagnosis of early Parkinson's at UCLA and had since made a radical switch to a vegetarian diet. Soon after, they sold their house and started an organic farm just north of Los Angeles. I was very glad to hear this news. I'm hopeful that the progression and severity of his disease will be positively influenced by his dietary and lifestyle changes.

## Alzheimer's Disease and Cognitive Decline

Alzheimer's disease is the leading cause of dementia in the elderly today. The numbers are astronomical: in 2017, an estimated fifty million people were suffering from it worldwide, and this number is expected to double every twenty years.[33] It's worth pointing out that premature and severe cognitive decline is in no way a normal part of aging, even though such assumptions are often made. While our recent increase in life expectancy, with more people living into their eighties and nineties, certainly contributes to the increase, our modern lifestyle and diet are likely to play a much bigger role. There are currently no proven therapies to prevent or slow the progression of Alzheimer's, drawing even more attention to the fact that scientists and doctors still have an incomplete understanding of its cause.

However, there are a growing number of studies linking this neurodegenerative illness with the gut. Many of the genes implicated in Alzheimer's suggest alterations in the functions of the immune system and a role of the gut microbiome in Alzheimer's development. Some of the most intriguing research comes from the Alzheimer's Disease Metabolomics Consortium at Duke University. Researchers led by Dr. Rima Kaddurah-Daouk identified a link between the liver, the gut microbiome, and biomarkers of neurodegeneration, causing them to propose a gut-liver-brain axis as part of the BGM network, implicated in the onset of the illness. "We can point now to problems in the gut and problems in the liver that are communicating with some aspects of disease in the brain in Alzheimer's," Kaddurah-Daouk observed, "suggesting we really should pay more attention to the interconnectedness of the brain with other organs."[34]

Analyses of 1,556 subjects from the Alzheimer's Disease Neuroimaging Initiative revealed that patients had reduced levels of primary bile acids in their blood and *increased* levels of certain secondary bile acids, which were found to be associated with poorer cognitive function, reduced brain glucose metabolism, and greater brain atrophy.[35] Primary bile acids are produced in the liver from cholesterol, stored in the gallbladder, emptied into the small intestine, and then reabsorbed in the gut in order to reenter systemic circulation; therefore, both primary and reabsorbed bile acids reach many organs in the body, including the brain. However, during their short transition through the gut, they interact with different groups of microbes, which modify their chemical properties and transform them into a group of *secondary* bile acids.[36]

Despite the positive role that many primary and some secondary bile acids play in our health, like aiding the absorption of fats in the small intestine, researchers have found that certain secondary bile acids have a potentially harmful effect on brain function. What's especially fascinating is that secondary bile acids are produced only by gut microbes possessing an enzyme called 7-alpha hydroxylase, which is essential for the conversion to occur. Without the abnormal function of these bile-acid-metabolizing gut microbes, there would be lower levels of harmful secondary bile acids throughout our bodies and brains. It remains to be determined if genetic alterations in bile-acid metabolism, altered relative abundances of bile-acid-metabolizing microbes, and/or dietary influences play a role. Significant research efforts are under way to answer these questions. In fact, Dr. Kaddurah-Daouk's research team found that increased secondary bile acids were not only linked to two biomarkers of neurodegeneration—amyloid and tau accumulation in the brain—but their levels in the brain also correlated with the progression of symptoms from mild cognitive decline to full-blown Alzheimer's disease. This stunning finding strongly suggests that the gut plays a pivotal role in the development of Alzheimer's.[37]

"We have studied the brain in isolation for too long," Dr. Kaddurah-Daouk concluded about her study, an elegant embodiment of network science. "Not only should we be targeting the brain; we should be targeting other organs that *talk* to the brain."

A few years ago, I had the pleasure of meeting Dr. Kaddurah-Daouk at a conference about the role of the microbiome in aging, organized by the National Institute on Aging (NIA) in Washington, DC. We'd both given presentations and seen each other speak; the dynamic potential in collaborating was immediately apparent. As a result, my research group was invited to be part of an international consortium led by Dr. Kaddurah-Daouk, Dr. Sarkis Mazmanian, professor of microbiology at the California Institute of Technology, and Dr. Rob Knight, professor of pediatrics and of computer science and engineering at the University of California in San Diego (and cofounder of the American Gut Project). Together we're aiming to identify how diet-induced changes in gut microbial metabolites and inflammation affect the brain, as well as assess the causal relationship between these metabolites and cognitive decline. Although it's a gargantuan effort to coordinate the research efforts of some thirty-five senior investigators from fifteen research institutions throughout the US and Europe, this study, having already gathered several thousand research subjects, is setting the gold standard for investigating the role of the gut microbiome in chronic brain disorders. Considering the magnitude and sophistication as well as the leadership of this consortium by the best in the field, it's likely that breakthroughs in the understanding of this devastating disease will be achieved in the next five to seven years, as well as established evidence for the benefits of dietary interventions.

## Autism Spectrum Disorder

Autism spectrum disorder, or ASD, is a devastating neurodevelopmental disorder affecting one in forty-five children in the United States. Of all the brain disorders, ASD has increased at the most dramatic rate, nearly tripling in incidence over the last decade and a half. About a half million people on the autism spectrum will become adults in the next decade, "a swelling tide," as a statement from the Centers for Disease Control and Prevention put it, "for which the country is not prepared."[38]

The exact causes of ASD remain unclear but, like all chronic brain disorders, it's believed to involve a combination of genetic and environmental risk factors. Given that the genetic risk of this disorder remains steady at 50 percent, the striking increase in ASD suggests that external influences like diet are prominent.

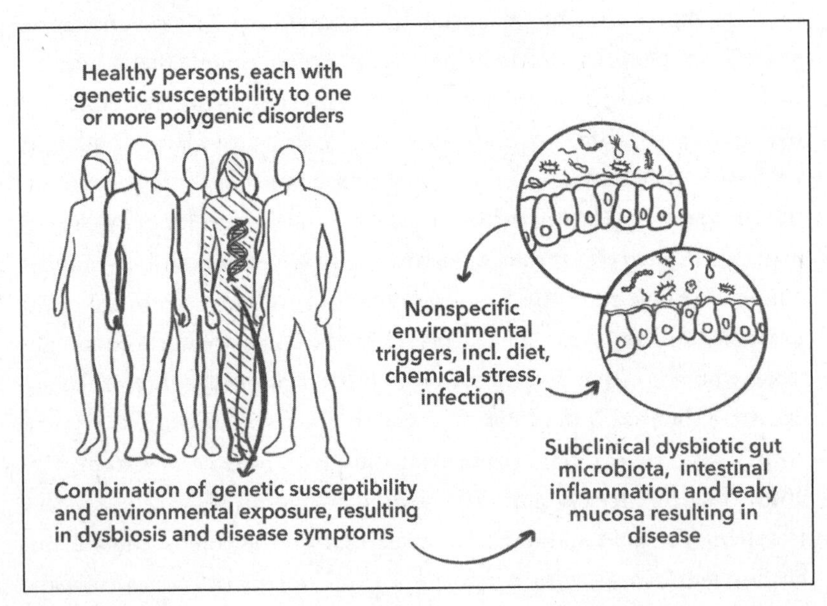

ASD is diagnosed based on the presence and severity of impaired social communication and repetitive behavior, but immune dysregulation and gastrointestinal issues are also commonly experienced by affected patients. And as with the neurodegenerative disorders, there's been a good deal of recent research reporting that alterations in the gut and its microbiome contribute to the symptom complex. Perhaps most significant, a growing number of studies have identified risk factors in pregnancy that can make the baby more likely to develop ASD or other neurodevelopmental disorders; these factors include maternal stress, infections, and age.[39] Numerous epidemiological, clinical, and animal studies have found that immune-system activation associated with infections and from poor metabolic health of pregnant mothers, plus the associated alterations in the microbiome, increase the risk for fetal development of ASD. For instance, one study modeled maternal immune activation in pregnant lab mice, which resulted in global changes in the composition of the adult offspring microbiome. The resulting imbalance of bacteria in the gut was correlated with lasting behavioral abnormalities, neuropathologies, immune dysfunction, and deficient gastrointestinal integrity.[40] Consequently, even though ASD afflicts a population ranging from infants to young adults, a major cause of its increased prevalence is likely to begin with the health of the mother. In the US, nearly 60 percent of women of childbearing age (twenty to thirty-nine years of age) are overweight, one-third are obese, and 16 percent have metabolic syndrome.[41] A 2012 study published in the journal *Pediatrics* showed that the risk of having an autistic child increased up to 2.4-fold when a mother had metabolic syndrome. Without putting an undue responsibility on pregnant mothers, I strongly feel it is important for both the public and doctors to be aware of these risks, which are all too often ignored. In the next chapter, I'll further explore the role of diet for pregnant mothers, their babies, and the risks for ASD.

Gut microbial alterations are associated with a compromised intestinal barrier and have also been found in children who suffer from ASD. As with Parkinson's, these patients exhibit a decreased Bacteroidetes-to-Firmicutes ratio, as well as increased *Lactobacillus* and *Desulfovibrio* species, all of which correlate with the severity of the disorder. This severity has been linked to a reduction in short-chain fatty acids, which have a protective effect on intestinal permeability. The relative abundance of *Prevotella* is also decreased in ASD patients (as in Parkinson's and Alzheimer's), as is consistently the case to some degree in industrialized societies eating a Western diet.

Therefore, Drs. Rosa Krajmalnik-Brown and Dae-Wook Kang from the Department of Soil, Water and Environmental Sciences, together with Dr. James B. Adams, president and founder of the Autism Nutrition Research Center at the Arizona State University, performed a study to examine whether industrialization-related gut-microbiome changes play a role in the increasing prevalence of ASD.[42] They compared the *Prevotella* depletions of autistic children living in the US to healthy children living in the developing world to determine whether the difference in the relative abundance of this organism was even greater when compared with non-autistic children living in the US. As expected, they found that American ASD patients appeared even *more* "Westernized," a situation the authors referred to as the hyper-Westernization of the fecal microbiota of autistic children. Such studies have led some researchers to believe that lack of beneficial gut microbiota—in particular the SCFA producer *Prevotella*—impairs neurological health.[43]

Dr. Krajmalnik-Brown and her collaborators took their findings even further by exploring a potentially revolutionary way to treat ASD. In order to evaluate the benefit of transferring gut microbes from a healthy individual to an ASD patient, they performed a small, open-label clinical trial on eighteen children with ASD. (An open-label trial is one in which both the researchers and the participants know who is getting the treatment being tested.) This study entailed treating the children with a fecal transplant, also referred to as microbiota-transfer therapy (MTT). This emerging treatment combines a two-week antibiotic treatment with a bowel cleanse and a stomach-acid suppressant, in order to greatly suppress the existing gut microbial ecosystem, followed by a fecal-microbiota transplant (FMT) from a healthy donor. The extended fecal-microbiota transplant was accomplished by using a high initial dose followed by daily and lower maintenance doses of the transplanted amount for seven to eight weeks.

Based on the patients' subjective ratings of their gastrointestinal (GI) symptoms, the investigators determined they were reduced by approximately 80 percent by the end of treatment, including significant improvements in constipation, diarrhea, indigestion, and abdominal pain. Most surprising, however, they found that the behavioral symptoms of the autistic children also improved significantly. Furthermore, all of these improvements remained at their eight-week follow-up appointments.[44]

In my opinion, the most amazing insight to come from this study is that it's possible to knock out the compromised ecosystem of the children's microbiomes and restore it with a healthier version with more bacterial diversity and an abundance of beneficial taxa, including *Bifidobacterium* and *Prevotella*. Unfortunately, other than in this particular study, numerous attempts to replace a bad gut microbiome with a healthy one have not been successful. Microbes injected into an intact microbial ecosystem, whether it's healthy or compromised, generally don't last long in their new environment. For example, in most people who take probiotics, the changes achieved are no longer detectable after forty-eight hours. For the majority of people, probiotics don't have persistent effects on gut microbial abundances or functions. Likewise, attempts to recolonize the gut with fecal microbial transplants to treat various conditions—such as irritable-bowel syndrome, inflammatory bowel disease, or obesity—have all failed in clinical studies. In general, this is due to the perpetual inclination of any ecosystem toward stability, resilience, and resistance—even if the return to its original state isn't beneficial. The very properties that prevent us from getting sick are also the ones that can resist even a healthy change.

However, this study of ASD children is an astounding exception to the rule. The researchers were not only able to achieve a successful colonization by donor microbiota, relieving both GI and ASD symptoms, but they also effected sustained changes and improvements. In fact, when the researchers consulted with their eighteen subjects again *two years* after the initial treatment, they discovered that most improvements in their subjects' GI difficulties were maintained, and ASD-related symptoms had improved even *more*.[45] The beneficial changes in gut microbiota remained, including significant increases in bacterial diversity and in relative abundances of *Bifidobacteria* and *Prevotella*. That said, it's important to emphasize that this work was uncontrolled, not placebo controlled, meaning that the participants and their desperate parents knew they were getting a potentially therapeutic intervention. Still, encouraged by these remarkable results, a glimpse of a possible therapy to treat children with ASD and GI symptoms, the researchers started a double-blind, placebo-controlled trial in adult ASD patients, and are planning a similar study in children as soon as pandemic-related restrictions on clinical research are lifted. "We are finding a very strong connection between the microbes that live in our intestines and signals that travel to the brain," concluded Dr. Krajmalnik-Brown. "Two years later, the children are doing even better, which is amazing."[46]

## The Reinforcing Effect of Stress in Chronic Brain Disorders

In all of these brain disorders—depression, Parkinson's, Alzheimer's, and ASD—affected persons' compromised ability to interact with the world in a healthy and adaptive way means that they experience a chronically activated stress response. With cognitive decline, for instance, there's anxiety associated with the realization, time and again, that their brain and memory is failing them. In ASD, impaired interpersonal and socializing skills and resulting loneliness cause stress. In depression, the interaction of genetic and epigenetic factors rendering persons hyperresponsive to stress makes them more vulnerable to anxiety disorders as children, and this in turn makes them more likely to develop depression later in life. In fact, some evidence suggests that the earlier anxiety begins for a person, the more likely it is that depression will follow.[47] Furthermore, in these disorders, the constant engagement of the fight-or-flight physiology and consequent conditioning of the fear response—consistently generating anxiety, sadness, and anger—creates an extra burden on the top-down input to the BGM network. This in turn influences the gut, causing adverse shifts in the microbiome, such as an increase in bacteria that produce the tryptophan metabolites kynurenine and indoles, and thereby increasing the low-grade immune activation. These shifts are communicated—by signals like metabolites, immune mediators, and vagus-nerve activity—back to the brain, reinforcing and perpetuating the original disorder and possibly even leading to structural—so-called neuroplastic—brain changes. The circular conversation goes around and around.

There are few areas in BGM research that show such consistent results from preclinical and human studies as those demonstrating the effect of acute and chronic stress on gut microbial composition. These findings have been meaningfully expanded by a 2017 study published in *Scientific Reports* by Ioana A. Marin and colleagues, showing that emotional behavior, microbiota composition, and the metabolite signature of the microbiome dramatically changed in chronically stressed mice.[48] Consistent with prior research on the effects of chronic stress on the microbiota, they observed a significantly reduced proportion of *Lactobacillus* and increased circulating kynurenine levels as the most prominent changes. Remarkably, giving the stressed mice a *Lactobacillus* probiotic restored intestinal *Lactobacillus* levels enough to reduce the kynurenine levels and improve behavioral abnormalities. In fact, members of the *Lactobacillus* genus have the capacity to produce high levels of hydrogen peroxide ($H_2O_2$) as a means of maintaining their niche within the highly competitive gut microbial ecosystem. The findings suggest that production of $H_2O_2$ by *Lactobacillus* may protect against the development of chronic stress–induced, depressionlike behavior by direct inhibition of the intestinal enzyme IDO. This in turn decreases kynurenine, which we know is associated with depression and other brain disorders.

These dramatic effects with a "psychobiotic" of *Lactobacillus* strains have not been reproduced in humans, but these experiments strongly support the possibility that other microbiome-targeted therapies, including diet, might be a useful component of antidepressant therapy. Furthermore, other still-evolving research for *all* of these disorders has shown that what we eat has an impact on their severity, giving us all an opportunity to improve brain health through diet.

**Chapter Five**

# HOW DIET REGULATES THE BRAIN-GUT-MICROBIOME NETWORK

In the Hippocratic Corpus, some sixty ancient Greek medical treatises collected over two thousand years ago, fasting is mentioned as a treatment for epilepsy. It wasn't until the early 1920s,[1] however, when researchers at Harvard Medical School made clinical use of fasting, that it became known as a treatment for seizure relief in patients with refractory epilepsy—that is, for the 30 percent of patients who are unresponsive to antiepileptic drugs. As fasting is associated with a switch of the metabolism to one of ketosis, over time, these empirical insights developed into a more specific recommendation to use the ketogenic, or keto, diet—a high-fat, high-protein, and low-carbohydrate regimen that forces the brain cells' metabolism to burn fats as a primary source of fuel (a state called ketogenesis).[2]

Although the short-term clinical benefits of a keto diet among refractory epilepsy patients have since been widely accepted, scientists didn't have a clear explanation for *why* it works—until recently. In 2018, Elaine Hsiao, my colleague at UCLA, identified the specific gut bacteria and some of their metabolites that mediate these anti-seizure effects. In Hsiao's study, published in the prestigious journal *Cell*, researchers worked with laboratory mice to identify two types of bacteria elevated by the keto diet: *Akkermansia muciniphila* and *Parabacteroides*.

The researchers also found levels of biochemicals in the gut and blood of the mice that were altered in ways that affect neurotransmitters in the hippocampus, a region of the brain that is involved in generating seizures.[3] The findings showed that these bacteria produced increased levels of the neurotransmitter gamma-aminobutyric acid (GABA), which suppresses the activity of nerve cells by the same mechanism as the psychoactive drug Valium. Both GABA and Valium can open a gate in the membrane of nerve cells that makes them less excitable to a variety of stimuli.

Though there's more work to be done to determine whether the same mechanisms identified by Dr. Hsiao in laboratory mice apply to humans, this research was the first to establish the involvement of the gut microbiome in the therapeutic effects of a specific dietary intervention for a serious brain disorder, and it beautifully captures the general direction in which the science of nutrition and the microbiome is headed.[4]

In the last several decades there's been an evolution in the types of studies that examine the link between another diet—the traditional Mediterranean diet—and its corrective effect on illness. Initially, there were broader comparative analyses, such as large-scale population studies (showing associations between a particular diet and health), epidemiological studies (looking at diseases in populations of people), and cohort studies (following subjects who share a defining characteristic over a long period of time). All of these showed an association between the Mediterranean diet and improved health. For example, a number of such studies demonstrated that an increased consumption of a large variety of fresh fruits and vegetables, nuts, seeds, and olive oil, as well as small amounts of lean animal protein primarily from poultry and fish, correlates with reported happiness as well as higher levels of mental health and well-being compared to those eating a typical Western diet.[5] However, as impressive as they are, these studies offer only associations; they don't provide evidence that a healthy diet actually *causes* better health or that an unhealthy one causes illness. Furthermore, in observational studies like these, there are inevitably other factors contributing to the differences—anything from more socializing among subjects on a particular diet to higher income to varying levels of stress, happiness, and physical activity—making it challenging to get a conclusive finding.

Nonetheless, over the last ten years, research has become more sophisticated, homing in on a more structured analysis of the connection between diet and health, as well as a more clinical view into the processes behind it. For example, the emerging field of nutritional psychiatry is an integral part of this movement, aiming to generate research that explores the impact of diet on mental health. In these more recent studies, researchers randomly assign participants to an experimental group or a control group and track what participants have actually eaten. Moving one step beyond these, the most recent studies have investigated the bottom-up contribution to brain disorders—messages sent from the gut and microbiome to the brain—pinpointing the ways in which specific microbes and their metabolites can influence a positive outcome. I consider these experiments to be the gold standard, offering scientific findings that demonstrate an unequivocal causal link between nutrition and mental health.

Though the traditional Mediterranean diet has been associated with improvement in many noninfectious illnesses—obesity, metabolic syndrome, cardiovascular disease, inflammatory bowel disease, and nonalcoholic fatty-liver disease, among them—I focus primarily on the three brain disorders discussed in the previous chapter. Truthfully, given the profound interconnectedness of all three diseases and the way in which all of our organs are affected in their development, I've come to view these illnesses as one complex, multifaceted syndrome. In my view, a separate diagnosis for each one serves only to artificially isolate them, causing each one to be treated by a different medical specialist with different drugs. Viewing these illnesses as one syndrome also seems apt, given that they're all associated with metabolic disturbances as well as chronic immune activation associated with an unhealthy diet. Perhaps the most crucial bond, however, is that research has shown them to be responsive, in one way or another, to the therapeutic impact of a largely plant-based diet, such as the traditional Mediterranean diet, which in contrast to vegan and vegetarian diets includes a small amount of animal products.

## Depression

A large number of observational studies have suggested that diet can affect a person's vulnerability to depression. We know that depression is determined by a complex interplay of genetic and epigenetic factors, including environmental, hormonal, immunological, and biochemical influences.[6] It would follow, then, that the food we eat, which influences all of these systems, would have a meaningful impact on the development and course of this debilitating mental-health disorder.

Recent research has shown that diets high in processed food, animal products, and refined sugars, such as the standard American diet, are associated with an increased risk of depression,[7] while diets based on vegetables, fruits, legumes, whole grains, and seeds, with a small amount of poultry and fish, are linked to reduced risk. Indeed, one meta-analysis examining the results of nine different studies on depression reported that adherence to such a diet yielded a 30 percent reduced risk of depression.[8] Although observational data always includes complicating factors like socioeconomic differences, the evidence these studies have brought forth is nevertheless impressive enough for me to point my patients who struggle with depression toward predominantly plant-based diets. I'm also increasingly confident in the data from newer studies using randomized, controlled dietary interventions, which can more rigorously examine the clinical benefits of such a diet.

One such study, from the Biomedical Research Centre Network on Obesity and Nutrition at the Institute of Health Carlos III, in Madrid, spanned eight years under the leadership of Dr. Almudena Sánchez-Villegas. It was the largest dietary-intervention trial to date designed to assess the effects of the traditional Mediterranean diet* on

---

* The **traditional Mediterranean diet**, which was the most popular diet in Italy, Greece, and Spain as late as the 1960s, is a *largely* plant-based diet characterized by a high intake of varied fruits and vegetables, olive oil, nuts, and cereals; a moderate intake of fish and poultry; a low intake of dairy products, red meat, and sweets; and a moderate amount of red wine consumed with meals.

Several traditional Asian diets have a similar composition. The **traditional Japanese diet** is rich in fish, other seafood, and plant-based foods, with minimal amounts of land-animal protein, added sugars, and fat. It consists of small dishes of simple,

cardiovascular-disease prevention.[9] However, this multicenter trial, known as the PREDIMED study (for PREvención con DIeta MEDiterránea), also performed a secondary analysis of the results to see if the Mediterranean diet also had beneficial effects on depression risk as compared with a control group on a low-fat diet. In a fitting confirmation of my "one syndrome" concept, researchers found that those following a Mediterranean diet benefited doubly, with a reduced risk for developing *both* cardiovascular events and depression.

For the primary study, in which researchers looked at the effects of diet on cardiovascular health, researchers enrolled 7,447 men and women aged fifty-five to eighty at high risk for cardiovascular illness and assigned them to one of three diets: a Mediterranean diet supplemented with extra-virgin olive oil, a Mediterranean diet supplemented with mixed nuts, or a control diet, in which the subjects were simply advised to reduce dietary fat. Participants were also given quarterly educational sessions about the Mediterranean diet and, depending on which group they were in, received free extra-virgin olive oil, mixed nuts, or small nonfood gifts. The primary goal was to determine the participants' risk for and rate of major cardiovascular events like heart attack, stroke, or death and to study how diet might affect those outcomes. The average risk for a major cardiovascular event was about 30 percent lower in both groups on the Mediterranean diet as compared to those on the control diet. The benefit was so significant, in fact, that this study had to be stopped midway for ethical reasons. The researchers couldn't in good conscience continue it.

In the subanalysis of this data, Dr. Sánchez-Villegas's team found a 20 percent reduction of depression risk in the group eating a

---

fresh, seasonal ingredients. The **traditional Okinawan diet** is based on green and yellow vegetables, especially sweet potato, and regular consumption of small amounts of small fish and pork. Modern Mediterranean and Asian diets have much higher proportions of red meat, dairy products, and highly processed foods, including sugars.

In contrast to the largely plant-based traditional diets, **vegan diets** are devoid of *all* animal products, including meat, seafood, eggs, and dairy. **Vegetarian diets** are free of meat, fish, and fowl flesh.

Mediterranean diet supplemented with nuts. While this is a clinically meaningful finding, it didn't reach statistical significance, meaning that the results might possibly be explained by chance. However, when the investigators analyzed only the subset of participants with a diagnosis of type 2 diabetes at enrollment, the benefit increased to 40 percent and did reach statistical significance.

Such favorable effects of the Mediterranean diet on both cardiovascular and mental health not only confirm the interconnectedness of these two chronic diseases, but also clearly indicate the vital part that diet can play in improving outcomes and preventing complications. I've seen such results in my own clinic. Time and again, when I treat my patients who have various types of brain disorders by putting them on a largely plant-based diet, I later learn that some of their secondary chronic conditions—such as diabetes, obesity, and fatty-liver disease—have also improved.

Since PREDIMED, two similar studies have had similar results: HELFIMED (HEaLthy eating For lIfe with a MEDiterranean-style diet) and SMILES (Supporting the Modification of lifestyle In Lowered Emotional States). The HELFIMED study, published in *Nutritional Neuroscience* in 2019, investigated whether a Mediterranean-style diet supplemented with fish oil improved the mental health of adults with self-reported depression.[10] The subjects were randomized and received food hampers every two weeks, Mediterranean-diet cooking workshops for three months, and fish-oil supplements for six months, while the control group attended social groups every two weeks for three months.

Likewise, the SMILES study, led by Felice Jacka, an associate professor of nutrition and epidemiological psychiatry at Deakin University in Australia and director of its Food and Mood Centre, investigated the effectiveness of dietary intervention in major depressive episodes.[11] Her research team theorized that teaching patients already receiving conventional therapies for moderate to severe depression how to change their eating habits to a Mediterranean-type diet would decrease their symptoms. The researchers also theorized that this method would be superior to a social support intervention, in which a member of the research team met one-on-one with participants and discussed subjects they enjoyed, without touching on emotionally charged issues.[12] The dietary intervention consisted of seven individual hourlong nutritional consultations with a clinical dietitian; the control subjects spent the same amount of time in social-support meetings. When the study finished after twelve weeks, thirty-one of the enrolled patients had completed the dietary intervention; twenty-five, the social-support-group treatment.

In both HELFIMED and SMILES, the dietary-intervention group showed significant improvement in their depressive symptoms after twelve weeks. In fact, in SMILES, 32 percent of patients experienced clinical remission from their major depression, compared to only 8 percent in the control group.

There are, of course, methodological limitations to these studies, such as the fact that the diet-intervention groups were aware of their treatment and presumably anticipated a benefit; such an "expectation bias" generally yields a placebo effect. Still, these three trials substantiate with greater authority the findings of previous epidemiological studies that eating a largely plant-based diet greatly reduces depression symptoms regardless of other therapies being used.

While the authors of these studies speculated that gut microbial changes brought on by the Mediterranean diet helped achieve the positive outcomes, the trials were not designed to explicitly investigate such a connection. Essential questions remain: Does consuming a traditional Mediterranean diet change the gut microbial composition and function in a way that improves mood? Is it possible to identify a specific molecular mechanism related to the change, or does the Mediterranean diet simply reduce low-grade immune-system activation in the gut and thereby the systemic inflammation and neuroinflammation associated with a Western diet?

To answer these questions, an Italian research group under the leadership of Drs. Danilo Ercolini and Paola Vitaglione from the University of Naples conducted a study to assess the effects of a Mediterranean diet on the relative abundances of gut microbes and their metabolites in overweight and obese people who were otherwise healthy.[13] Although the participants weren't selected for depressive symptoms, they experienced rapid and dramatic changes in the gut microbiome in response to the dietary intervention, and these results can be extrapolated to better understand the effects of this type of diet on depression.

The eight-week-long trial was done with eighty-two subjects. Participants were divided into two groups. One was given a personalized diet rich in fruit, vegetables, whole-grain cereal products, legumes, fish, and nuts. The control group ate a standard Western diet. Both groups consumed the same number of calories and the same ratio of macronutrients (carbohydrates, protein, and fat). In addition to standard blood tests, participants underwent a detailed analysis of microbial metabolites in blood, stool, and urine. Researchers also used a new analytical approach, a recent addition to the "-omics revolution" called foodomics, which analyzes the pattern of metabolites broken down from different components of food. This method allows for far more accuracy than self-reported data from participants, which is notoriously unreliable.

Researchers found that the group eating a Mediterranean diet experienced an increase in gut microbial gene richness, a measure of gut microbial diversity, which was inversely related to measures of systemic inflammation. They also observed an increase in the proportion of beneficial microbes like *Faecalibacterium prausnitzi*, which break down fiber into short-chain fatty acids and other metabolites, as well as a decrease in mucus-degrading microbes like *Ruminococcus*. As discussed in chapter 4, it's likely that other neuroactive metabolites—such as GABA or certain tryptophan metabolites—play additional specific parts in modulating the brain to decrease symptoms of depression, but based on these recent studies, I personally believe that the reduction of chronic systemic immune activation (metabolic endotoxemia) and the resulting reduced activation of immune cells in the brain are central to the natural antidepressant effect of this diet in many patients. Even though I doubt that nutritional psychiatry will *replace* antidepressants or cognitive behavioral therapy any time soon, the evidence from these studies argues for making diet an essential pillar in the treatment and prevention of chronic depression.

## Using Probiotics to Treat Depression

Mary, a fifty-two-year-old lawyer, came to my office complaining not only of depression but also of severe chronic abdominal pain and constipation. Clearly in considerable distress, she held both of her hands tightly over her belly as she sat down to discuss her problems.

"About six months ago, I started to have serious belly pain," she explained, eyes wide with concern. "And I've also really been struggling with depression. I've seen a number of doctors. But they couldn't help me. Recently, I read a book about psychobiotics being able to help with mood, and I wanted to get your opinion about which one I should take to help with my depression."

I explained to Mary that many of the recent articles she'd likely read about psychobiotics—live bacteria meant to confer a mental-health benefit by affecting the gut microbiota[14]—are not based on rigorous scientific evidence. Even so, such misinformation is consistently delivered by the media, bestselling books, and the Internet, with outsize promises that a certain bacterial supplement will improve mood, enhance cognitive function and prevent its decline, and even provide relief from epilepsy, attention-deficit hyperactivity disorder, and autism.

I suggested to Mary that before we discuss psychobiotics, we take a closer look at her symptoms. Before coming to me, she explained, she'd undergone extensive diagnostic evaluations for her pain, including endoscopies of her upper and lower GI tracts, a CT scan of her abdomen, and repeated blood tests. As in the great majority of patients that come to see me for similar symptoms, none of these tests had revealed any abnormalities. A previous doctor had also prescribed laxatives, which relieved her constipation only temporarily. Simultaneously, her primary-care physician referred her to a psychiatrist, who prescribed several antidepressant medications, including SSRIs, but Mary was unable to tolerate any of them, as they affected her concentration, sleep, and bowel movements. She'd experienced a similar sensitivity with other medications in the past.

After reviewing her lab results, I asked Mary more-general questions about her life. Her personal story revealed several important details that helped me to better understand her symptoms. First, when I asked whether she'd had any gastrointestinal symptoms before, she said she'd struggled with them throughout her life. In fact, she'd had months-long periods of significant abdominal distress and constipation since she was a teenager. She'd also struggled with episodes of anxiety and depression as long as she could remember, but especially in the last few years.

When I gently inquired about her more recent episodes, wondering if she could point to anything in particular that might have provoked them, she said she'd gotten a divorce several years ago, leaving her as a single parent raising her teenage son. Mary also opened up about the loss of her mother, who'd died four months prior, after a prolonged struggle with ovarian cancer. Though the potential link between such profound losses and Mary's medical symptoms was immediately clear to me, as is often the case with a patient in the midst of trauma or grief (and physical pain), she hadn't been able to step back far enough to make the connection. She'd only ever considered that her personal losses had led to her depression, despite the fact that her slide downward in mood had gone hand in hand with her gastrointestinal problems.

To help her see the relationship between the two more clearly, I explained the intricate connection between the mind and the gut, and the way certain behavioral factors, like the trauma she'd experienced, can disrupt the balance of this system, leading to both mood *and* GI tract changes. Though this explanation made good sense to her and offered a medical explanation of her symptoms, Mary still wanted to know whether a probiotic supplement might provide relief from her suffering. She'd already tried several probiotics for her constipation but she hadn't noticed any change in mood. I said a supplement might benefit the gut in some patients, but a significant benefit on her mind was less likely.

"Still, I would like to do something proactive," she replied.

I understand how tempting it is to believe there's a pill that will fix everything. Particularly with depression, the uphill trudge of putting one foot in front of the other toward a better day is exhausting. But truthfully, I don't think a probiotic will ever be a primary treatment for depression. However, as I told Mary, the good news is there *are* other interventions within her control—diet and lifestyle changes that undoubtedly benefit both the brain and the gut. Altering the communication between these two organs can effect a transformation, mentally and physically. But this takes commitment, certainly more energy and time than simply taking a probiotic alone.

Mary nodded, but she had done her homework. "Yes, but what about the research that shows that probiotics have been linked to a decrease in depression in patients?"

There *are* studies, I conceded, that show positive results with pre- and probiotics—conducted to identify, as with the earlier depression studies, whether there's a causal relationship between the gut microbiome and depression.

In one such study by researchers from the Tehran University of Medical Sciences, 110 depressed patients were randomly assigned to receive either a mix of two probiotics, *Lactobacillus helveticus* and *Bifidobacterium longum*; or galacto-oligosaccharide, a prebiotic; or a placebo for eight weeks.[15] The study's aim was to compare the effect of these two microbiome-targeted interventions on depression. The researchers found that taking probiotic supplements resulted in a significant decrease in symptoms compared to the other two groups.

In another placebo-controlled study, led by Dr. Rebecca Slykerman at the University of Auckland in New Zealand, 212 women who showed signs of depression and anxiety during pregnancy and the postpartum period were given the probiotic *Lactobacillus rhamnosus* HN001 with positive results.[16] The mothers in the probiotic treatment group reported significantly lower depression and anxiety scores than those in the control group. However, it is important to note that observed effects on mood were a secondary outcome; the primary outcome was to establish a positive effect of this probiotic on babies with eczema.

However, based on my own personal clinical experience with hundreds of patients suffering from digestive symptoms with depression and other mood disorders, as well as conversations I've had with colleagues and psychiatrists, I remain highly skeptical that currently available probiotics alone can have a clinically meaningful effect. There are numerous studies claiming health benefits for various diseases, but publication bias—meaning only positive results are reported—and limitations in study design also explain the absence of such effects in clinical practice. A similar conclusion was recently reached by the American Gastroenterological Association (AGA) in its clinical practice guidelines about the usefulness of probiotics in the management of gastrointestinal disorders. The AGA concluded, based on an in-depth review of the published literature, that probiotics have little, if any, evidence-based value in treating digestive diseases, such as irritable-bowel syndrome or inflammatory bowel disease.

In my experience, though, if a patient eats a standard American diet high in sugar, fat, and red meat and doesn't exercise but takes a probiotic supplement, the net effect is essentially the same as taking an expensive placebo. We simply can't rely on probiotics to replace a healthy diet and lifestyle. Most important, there are so many aspects of a plant-based diet that you could never replace with a pre- or probiotic pill: the tens of thousands of molecules in plants, each with its own small synergistic health benefit, which offer an overall—and long-term—health boost.

Rather than recommending a specific psychobiotic for Mary, my treatment plan instead addressed both sides of her brain-gut network. I recommended she start eating a primarily plant-based diet, rich in different types of fiber that fuel microbe diversity and richness, as well as incorporating various naturally fermented foods. This combination is intended to improve the gut microbial ecosystem *and* reduce systemic immune activation. I also suggested that she start regularly drinking green tea, as it has been shown in several studies to have antidepressant effects. This may be related to one of the main ingredients in it—essentially a polyphenol extract—which, as I'll discuss in chapter 7, has been shown to attenuate the brain's stress response. Last, I referred Mary to an integrative health coach in our clinic who could help her deal with her recent losses and teach her abdominal breathing and other mindfulness stress-reduction practices easy to incorporate into one's daily activities. The combination of such mind-directed therapies with a traditional Mediterranean diet, supplemented by naturally fermented foods, has proven to be quite successful in improving symptoms of depression. When I saw Mary months later during a follow-up visit, she thanked me for guiding her toward a more holistic treatment of her brain-gut-microbiome imbalance. Since we'd last met, she'd had a short course of cognitive behavioral therapy and had switched to a Mediterranean diet. She felt she had made good progress with working through her emotional difficulties, and even though she still had occasional bouts of dysphoria, overall she felt, by her estimate, 75 percent better.

## Autism Spectrum Disorder

Diet has always been considered to play a role in autism spectrum disorder, in terms of both the distinct dietary preferences of affected children and the attempts of desperate parents to find a regimen that will relieve behavioral and gastrointestinal symptoms. Unfortunately, with some exceptions, little progress has been made in the search for a dietary treatment. However, recent research focusing on pregnant women and the transgenerational effect of their health and nutrition has offered valuable insight. Systemic immune-system activation during pregnancy has been associated with an increase of autismlike behavior in mice as well as an increased incidence of autism in children.[17] Several studies have since investigated the ways in which maternal diet may foster low-grade immune activation and raise the risk of autism.[18]

Remarkably, two such studies found that even in the absence of maternal obesity, the consumption of a high-fat diet during pregnancy significantly increased the risk for autism and other mental disorders in babies. One animal study, from Professor Kjersti Aagaard's research group in the Department of Obstetrics and Gynecology at the Baylor College of Medicine, showed that a high-fat maternal diet, but not obesity per se, influences the gut microbial ecosystem of the mother.[19] The resultant microbial changes in offspring were only partially corrected by a low-fat diet after the animals were weaned. In addition, early exposure to this high-fat diet unexpectedly diminished the abundance of a strain of nonpathogenic *Campylobacter* bacteria in the offspring's gut, further supporting the concept that a mother's diet shapes the commensal microbial communities of her baby.

In 2016, similar findings were reported by Shelly Buffington, a postdoctoral fellow, researcher, and study coauthor, along with Mauro Costa-Mattioli, associate professor of neuroscience and director of the Memory and Brain Research Center at the Baylor College of Medicine in Houston. They showed that a high-fat diet, equivalent to eating fast food multiple times a day, not only spurred maternal obesity in mice but also altered their offspring's gut microbiome and caused social behavioral deficits, such as spending little time with their peers and seldom initiating interaction.[20] These social deficits were linked to an alteration in signaling within the brain's reward system. Subsequent fecal transplant experiments showed that an unbalanced combination of microbes in the mice born to the mothers on high-fat diets was responsible for the social deficits. When the researchers used genome sequencing, they discovered that one species, *Lactobacillus reuteri*, was reduced more than ninefold in offspring. The researchers decided to try to restore it, with remarkable results.

"We cultured a strain of *L. reuteri* originally isolated from human breast milk and introduced it into the drinking water of the high-fat-diet offspring," Buffington explained. "We found that treatment with this single bacterial strain was able to rescue their social behavior." Though other ASD-related behaviors, like anxiety, were not reduced by reconstituting this bacterium, the researchers did find that *L. reuteri* also bolstered the production of oxytocin, known as the bonding hormone, which helps to guide social behavior and when deficient has been associated with a diagnosis of ASD in humans. The findings demonstrate the gut microbiome's influence in regulating social behavior in laboratory mice, but it remains to be seen if such treatments would be effective in treating children with ASD.

Because of these encouraging results in mice, some have proposed antibiotics, probiotics, prebiotics, and fecal microbiota transplants to treat ASD. An open-label study in children with ASD found that eight weeks of treatment with oral vancomycin—a non-absorbable antibiotic that acts only in the gut—led to major improvements in both GI and ASD symptoms, although the benefits were lost within a few weeks after treatment.[21] Probiotics have also had mixed clinical results without long-term follow-up.

Before desperate parents are given false hopes of easy treatments with a magic psychobiotic, more work needs to be done to prove that these findings are applicable to ASD patients. However, this is a promising avenue of investigation for the eventual treatment of neurodevelopmental disorders.

## Cognitive Decline

Given their intertwined nature, the illnesses that make up metabolic syndrome—hypertension, heart disease, type 2 diabetes, obesity, and high blood cholesterol and lipid levels—are also the most vigorously studied risk factors for the development of premature cognitive decline. Each of these illnesses makes it more likely that a person will develop the others, but each one also increases one's risk of accelerated cognitive decline and Alzheimer's disease. Therefore, recent observational studies have been conducted and have confirmed that diet significantly affects one's risk for Alzheimer's.

A 2015 study, led by Deakin University professor Felice Jacka of the SMILES trial, found that low intake of nutrient-dense foods (e.g., salmon, kale, shellfish, and blueberries) and a high intake of Western processed foods (characterized as "roast meat, sausages, hamburgers, steak, chips, crisps, and soft drinks") can reduce the size of the left hippocampus, a brain region crucial to memory function that's repeatedly been found to shrink in the brains of Alzheimer's patients.[22] The researchers studied a group of 225 individuals in their early sixties without a diagnosis of Alzheimer's. They gave them two brain scans, four years apart. Throughout, they assessed their dietary patterns using a food-frequency questionnaire. A healthy dietary pattern with regular intake of plant-based foods was associated with a larger left hippocampus, but a Western diet was linked with smaller hippocampal volume. These results are consistent with observations previously made in animal studies. Unfortunately, as cognitive testing wasn't performed throughout the four-year period, and as the study design was observational, researchers were only able to speculate that the difference in diets was actually responsible for the brain changes and the progression of cognitive decline.

Several hybrid Mediterranean diets have been developed for specific disorders, such as DASH (Dietary Approaches to Stop Hypertension)[23] and MIND (Mediterranean-DASH Intervention for Neurodegenerative Delay). All are largely plant-based. DASH incorporates more heart-healthy fat. MIND combines the Mediterranean and DASH diets, emphasizing those aspects associated with improved brain function, such as polyphenol-rich berries and green leafy vegetables, but in contrast to the DASH and Mediterranean diets, it doesn't recommend high consumption of fruit, dairy, potatoes, or eating more than one meal of fish a week. The beneficial effects of the Mediterranean and DASH diets have been demonstrated in randomized intervention trials for the various manifestations of metabolic syndrome, including obesity, high blood pressure, diabetes, high blood cholesterol and lipid levels, insulin sensitivity, metabolic endotoxemia, depression, and cognitive decline—all of which are linked to one another and also to Alzheimer's disease.

The MIND diet was created in 2015 by the late Dr. Martha Morris, a professor in the Department of Internal Medicine, assistant provost of community research, and the director of the Rush Institute for Healthy Aging. The diet is based on her own pioneering research into preventing Alzheimer's disease through nutrition.[24] She tested the MIND diet among some 970 participants of the Rush Memory and Aging Project, volunteers living in retirement communities and senior public-housing units in the Chicago area. The participants underwent annual neurological evaluations and dietary assessments for nine years. The main goal was to see if the degree of adherence to the MIND diet would be associated with cognitive performance and the development of Alzheimer's. Researchers devised the "MIND diet score" to assess how well participants followed the diet and to track changes in cognitive function. And indeed, those who had the highest commitment showed significantly slower decline (as assessed by a cognitive test score) and lower rates of Alzheimer's during the study period. In fact, the effect was a stunning 53 percent reduction in the rate of Alzheimer's for persons in the highest third of the MIND scores and a 35 percent reduction for the middle third when compared with the lowest third of scores. As the data analysis didn't reveal any statistical evidence that the association between MIND diet scores and Alzheimer's incidence was mediated by obesity, metabolic dysfunction, or cardiovascular disease in the participants, the researchers concluded that adherence to the MIND diet was likely to have had a direct effect on the brain.

Nutrients absorbed in the small intestine and metabolites generated by gut microbiota, both from a largely plant-based diet, are thought to be key components of this improvement. Many animal studies show that gut microbiota are negatively influenced by a diet high in animal fat, causing neuroinflammation, decreased memory, increased anxiety, and a reduction in brain-derived neurotrophic factor (BDNF), a fundamental nerve-growth protein essential for learning and long-term memory in the central nervous system.[25] In addition, studies have shown that the Mediterranean diet is associated with increased abundances of beneficial bacterial genera like *Lactobacillus*, *Bifidobacterium*, and *Prevotella* and with a simultaneous decrease in the pathogenic *Clostridium*. Overall, these diet-related gut microbial changes result in a microbiota profile linked to several metabolic health benefits, such as lower levels of bad cholesterol and blood lipids and a reduction in systemic immune activation.[26]

Inspired by the University of Naples study mentioned above, of the gut microbiomes of overweight people, Paul W. O'Toole, professor of microbial genomics at University College Cork, Ireland, along with a consortium of researchers from five different European countries, aimed to look at the effects of the Mediterranean diet on the gut microbiome in relation to cognitive decline and other measures of frailty.[27] In this context, *frailty* refers to the development of chronic low-grade inflammation, loss of muscle and bone mass, cognitive-function decline, and increased risk of type 2 diabetes, Alzheimer's, or Parkinson's, all common among the elderly in developed countries.

The researchers assessed the effect of a yearlong dietary intervention on the gut microbial ecosystem and associated symptoms in 612 people of ages sixty-five to seventy-nine. Of these, 323 were on the Mediterranean diet, while the 289 control subjects continued with their normal diet. This study found that those on the Mediterranean diet showed a significant increase in the number and function of gut bacteria associated with better cognitive function, along with decreased inflammatory markers in the blood and lower frailty.

The specific microbes that became more abundant on the Mediterranean diet—the "diet-positive" taxa—included *Faecalibacterium prausnitzi*, *Roseburia*, *Bacteroides*, and *Prevotella*, all known for their association with metabolic health. A majority of these diet-positive taxa had already been linked to health-promoting activities, including production of short-chain fatty acids and anti-inflammatory molecules, as well as negative associations with diseases like type 2 diabetes and colorectal cancer.

Some of the same health-promoting microbes are consistently found in the traditional hunter-gatherer populations discussed in chapter 3; they reflect a diet abundant in unprocessed, fiber-rich foods and devoid of the chemicals added to processed foods. They are also associated with an increase in the consumption of the complex carbohydrates, or fiber molecules, prominent in the plant-based Mediterranean diet. It's interesting that an increase in similar taxa was also observed in the University of Naples study, which was performed in a much younger population, suggesting that these beneficial dietary effects are not limited to the elderly but likely benefit people of all ages. In fact, the two research teams have since collaborated, devising a study that links the Mediterranean diet to a rise in certain short-chain fatty acids associated with a reduced risk of inflammatory diseases, diabetes, and cardiovascular disease. Most important, however, this study demonstrated that a Mediterranean diet, even when adhered to for only a year, is strongly correlated with reduced frailty, improved cognitive function, and reduced inflammatory markers in the blood.

By contrast, microbial genera that showed a *decrease* in abundance on the Mediterranean diet included *Ruminococcus*, *Coprococcus*, and *Veillonella*, all known to be more abundant in the gut of people consuming a typical unhealthy Western diet containing a large amount of simple carbohydrates, or sugars.

When the authors looked beyond the relative abundance of good and bad microbes and evaluated the diet-induced changes in their *functional* metabolic profiles, they identified dramatic differences throughout the microbiome-response "landscape." Greater numbers of diet-positive microbes were associated with an increase in the microbial consumption of the complex, nonstarchy carbohydrates that make up a large portion of the Mediterranean diet. In contrast, a reduced number was associated with an increase in microbial simple-sugar consumption, the refined sugars that make up a significant part of the Western diet. A negative microbiome response was also accompanied by an increase in the microbial production of several secondary bile acids, the same ones that Duke University's Dr. Rima Kaddurah-Daouk showed to be associated with adverse brain changes and cognitive decline, implicating them in the development of Alzheimer's disease.[28]

However, one of the most intriguing findings from this study is that the two groups of microbes not only responded differently to the Mediterranean-diet intervention, but that they played very different roles within the gut microbial network. When applying the same mathematical approach that has been used to characterize other complex systems, such as the brain, as described in chapter 2, the microbes that increased on a Mediterranean diet occupied central and influential positions in this network. In network-science terms, this means that these microbes have control over all the other microbes and their functions within the network. On the other hand, microbes that decreased on the diet intervention occupied less influential, peripheral positions in the network. What may sound like an esoteric finding actually has major implications for a new understanding of the benefits of a largely plant-based diet. The location and influence of the diet-positive taxa point to their importance for the stability of the entire microbial ecosystem, making them "keystone" species. The loss of keystone species in any ecosystem—such as the wolves in Yellowstone or the bison on the prairies—has a profound impact on the whole ecosystem's health.

While we have known the beneficial effects of a largely plant-based diet on certain microbial functions and on gut health for a while, these novel findings reveal an even bigger benefit. They give us a clearer understanding of a healthy diet's ability to promote the resilience of our brain-gut-microbiome ecosystem in the face of stress and thus the overall health of our body.

A growing number of microbiome scientists are applying network science and graph theory to the gut microbial networks. My own research group has begun to link the network characteristics (such as centrality, hubs, resilience) of the gut microbiome with the network characteristics of the brain, applying this "multiomics" systems approach to the BGM network in health and disease. As I explained in chapter 2, it doesn't matter if a network is made up of billions of nerve cells in the brain or trillions of microbes in the gut. The rules governing the functions of *all* such systems are very similar. The characterization of biological interconnectedness is critical, in fact, to understanding the interplay between our own health, what we eat, and how we interact with the world.

In the next chapter, I will continue to explore interconnectedness of a different sort—the critical relationship among exercise, mental health, and a healthy diet. A plant-based diet isn't the only component of a long and fruitful life. Recent research suggests that diet and exercise interact to have an even more beneficial effect on our health.

**Chapter Six**

# A BROADER CONNECTION

## How Exercise and Sleep Affect Our Microbiome

The human body is a closely interconnected network in which the brain, the gut, and the microbiome are major hubs. If a mismatch is created in this brain-body network, it can lead to disruptions that manifest as chronic low-grade inflammation and increased risk of chronic disease. While diet is one of the most important strategies we have for reducing this risk, the science is clear that our lifestyle patterns, especially exercise and rest, also greatly influence our well-being, including the composition and function of our microbiome. While some of the variables that influence our health are out of our control—such as genetic vulnerabilities and socioeconomic circumstances—there are interventions that give us a hand in our own fate.

### Exercise

We've known for decades that physical exercise is one of the pillars of health and longevity. The benefits of regular exercise on metabolism and cardiovascular fitness—such as lowered risk for heart attacks and strokes, improved brain health, reduced depression and anxiety, and reduced cognitive decline—have all been well documented. Conversely, a sedentary lifestyle is a critical contributing factor to the high rates of disease in our current health crisis. Recent studies have also found that exercise can add healthy years to our lives. A 2020 Harvard T. H. Chan School of Public Health study led by Dr. Frank P. Hu and Frederick J. Stare demonstrated that at least thirty minutes of moderate to vigorous physical activity per day was one of five lifestyle traits that can increase the number of disease-free years that one can add to one's life expectancy. The others were healthy diet, normal body weight, no cigarette smoking, and moderate alcohol consumption. The authors showed that adhering to these five simple guidelines can add seven to ten extra disease-free years when adopted by the age of fifty.[1] A longer life without dependency on the medical system with its growing list of medications is attainable even when healthy habits are forged in late middle age.

Similar findings were reported by a group of Finnish investigators from the University of Helsinki led by Dr. Solja Nyberg. In a prospective, multi-cohort study of 116,043 participants from several European countries, a statistically significant association was found between many of the same healthy lifestyle choices and an increased number of disease-free years.[2] Researchers found that the factors related to the greatest illness-free years were physical activity, a healthy body-mass index, no smoking history, and moderate alcohol consumption. Several of these elements were also linked to a prolonged life without type 2 diabetes, cardiovascular and respiratory diseases, or cancer. While neither the Harvard nor the University of Helsinki studies *demonstrated* a causal role of exercise or any of the other factors, they both make a strong case for one.

Not only is exercise beneficial in much the same way that a healthy diet is, but the Harvard study also suggested that there may be a positive interaction *between* diet and exercise. That is, the combination of a healthy diet and exercising every day had a greater positive effect on healthy life extension than either one by itself. I believe together they have a synergistic effect on the health of the BGM network, preventing the maladaptive engagement of the immune system. Conversely, a lack of exercise and a poor diet cause the low-grade systemic immune activation that results from abnormal communication between the gut microbiome and the gut-based immune system. Consistent with the circular communication in this network, evidence has emerged that exercise has a beneficial effect on gut microbiota and improves exercise performance.

As with much of microbiome science, the first evidence came from studies conducted on lab rats. Rats allowed to run around freely were found to have gut microbiota different from their counterparts whose activity was restricted, as well as an increased level of butyrate, a short-chain fatty acid.[3] SCFAs are produced by certain microbes that ferment dietary fiber in the colon. The most common of these are butyrate, acetate, and propionate, which have positive effects on the gut, immune system, and brain—bolstering the gut wall, normalizing immune function, and eliciting a feeling of satiety.

These early observations about gut microbiota changing in more physically active rats were followed by a pivotal study of elite Irish rugby players performed by a group of investigators from the Alimentary Pharmabiotic Centre (APC) at University College Cork, Ireland, under the leadership of Professor Fergus Shanahan.[4] The researchers compared several gut-microbiome traits, as well as indicators of muscle activity and low-grade immune activation in the blood, between rugby players and a healthy control population composed of people with normal to elevated BMI but leading more sedentary lives. Researchers found significant differences between the groups in gut microbial diversity and relative abundance of organisms, as well as in activity of metabolic pathways and fecal metabolites. The athletes had greater microbial diversity and richness, as well as a greater abundance of *Akkermansia*, well-established as beneficial for gut health, and several other SCFA-producing taxa. These microbial changes were also associated with lower measures of systemic immune activation and higher levels of creatine kinase, an enzyme that varies with the amount of muscle activity. In addition, the athletes had more of the gut microbial genes required to generate the SCFAs acetate, butyrate, and propionate, as well as those needed for amino-acid and carbohydrate metabolism. These increases were linked with better fitness and overall health. However, because this study didn't control for the fact that the rugby players were eating a diet higher in protein and calories, the researchers couldn't tell if the differences were also influenced by what the athletes were eating.

A longitudinal study since performed in healthy humans demonstrates that, independent of dietary changes, endurance exercise does indeed have an effect on the composition and function of the gut microbiome. This study, done by a team from the Department of Kinesiology and Community Health at the University of Illinois at Urbana-Champaign under the leadership of Dr. Jeffrey Woods, explored the impact of six weeks of endurance exercise on the composition and function of the gut microbiota in both lean and obese adults, while also controlling for diet.[5] Researchers gathered eighteen lean and fourteen obese subjects, all of whom were living a largely sedentary lifestyle. Participants took part in a six-week, three-day-a-week supervised program of endurance-based exercise that progressed from thirty to sixty minutes per day and from moderate to vigorous intensity. Afterward, the subjects returned to their original sedentary lifestyles for another six weeks. Fecal samples were collected before and after the six weeks of exercise and six weeks after the return to inactivity. The exercise program resulted in significant changes in body composition, with increased total lean body mass and a reduction in the relative proportion of body fat. Furthermore, these changes were associated with exercise-induced increases in gut-health-promoting SCFAs. This beneficial effect was demonstrated at multiple levels of investigation: by increases in microbes capable of SCFA production (including the order Clostridiales and the genera *Roseburia*, *Lachnospira*, and *Faecalibacterium*), in genes associated with microbial SCFA production, and in fecal SCFA concentrations assessed by metabolomics, a technique which quantitates microbial metabolites. The investigators found that these changes in gut-microbiota diversity were not uniform in all participants but depended on the participant's body mass index or BMI. An exercise-induced increase in fecal concentrations of SCFAs was observed primarily in lean participants and only to a lesser degree in obese ones. The leaner subjects on a regular exercise program benefited the most in terms of gut health.

It's not surprising that these changes were largely reversed once the exercise program ceased. The authors concluded that exercise generates compositional and functional changes in the human gut microbiota, dependent on obesity status but independent of diet and contingent on the continuation of regular exercise. For workouts to benefit the microbiome, they must be regular. Although this study didn't directly address the point, it's plausible that the changes in gut microbial metabolites induced by vigorous exercise help produce the increased sense of well-being—the "runner's high"—that often accompanies it.

In contrast to the indisputable evidence supporting the health benefit of regular, moderate exercise, *extreme* exercise has been found to be problematic for gut health and overall well-being. I remember David, a thirty-seven-year-old runner, who came to my office a few years ago with an illustrative complaint: a recurring case of diarrhea that for the previous two years had been kicking in around the twenty-mile mark. It reliably kept him from reaching the finish line of every marathon he ran in. David was eager to figure out what was causing such an unnerving problem and what he could do about it. He'd recently read an article in a runners' magazine that suggested there might be a connection between his recurring diarrhea and dysregulation of his gut microbiome.

As it happened, I'd recently been invited to speak at the annual meeting of the American College of Sports Medicine, in Denver, where I'd learned more about the detrimental effects extreme exercise can have on the gut microbiome and connectome. I told David about the results of a recent study led by J. Philip Karl at the Military Nutrition Division of the US Army Research Institute of Environmental Medicine, in Natick, Massachusetts.[6] The team's aim was to examine whether high-intensity endurance exercise could have negative consequences on gut microbial composition and metabolic activity and whether these effects were related to a change in intestinal permeability—"leaky gut." In this study, seventy-three soldiers were provided three rations of food per day and could choose to add protein- or carbohydrate-based supplements during a four-day cross-country ski march. Intestinal permeability, blood samples, and stool samples were measured before and after the strenuous excursion. The observed changes varied, but the average permeability of the gut increased by 60 percent and was associated with an increase in systemic immune markers. The exercise-induced changes in gut microbial composition included a decrease in anti-inflammatory genera, such as *Bacteroides*, *Faecalibacterium*, and *Roseburia*, and an increase in the relative abundance of several rare, harmful taxa. These changes, along with a reduction in several stool metabolites, including the amino acids arginine and cysteine, were associated with the increased gut permeability.

"But I thought exercise was supposed to be good for the gut," David said. I told him that in principle he was correct but explained that there's a difference between extreme exercise and moderate exercise, like going to the gym or daily jogging.[7] Gastrointestinal symptoms—including bloating, cramps, diarrhea, heartburn, nausea, vomiting, and bloody stools—are reported by about 20 to 50 percent of extreme athletes,[8] more commonly by females.

"Overall, the majority of these athletes do not experience such adverse symptoms," I explained. "The reason you're getting sick, where other endurance athletes might not, has to do with the resilience of your gut to the physical stress of long-distance running, and this difference is related to differences in the gut microbiome."

"But how do my microbes even know how much I exercise?" David asked.

Good question. I've wondered about that myself. Does the body have a specialized signaling system that informs the hundred trillion microorganisms in our gut whether we're behaving like couch potatoes or exercising obsessively?

Here's what we do know: physical exercise activates the autonomic nervous system; its signals to the gut can change peristalsis, regional transit, secretion of fluid and mucus, intestinal blood flow, and intestinal permeability. These effects change the microbes' habitat, and they adjust—to a degree. Extreme athleticism, despite the excitement and sense of accomplishment that can come with it, may create a mismatch like the one between our modern daily challenges and our ancient stress-response system. The demands made of the body by high-intensity endurance exercise—an ultramarathon, a triathlon, or military boot camp—can ring alarm bells in the brain, creating an exaggerated stress response. In some vulnerable individuals, these increased stress signals can lead to leaky gut and immune-system activation, with all of its accompanying negative effects on the body and brain, as well as changes in gut microbial abundances and behavior.

My advice to David was to switch to a gut microbiome–supporting diet. He needed to counteract his extreme exercise–induced reductions in SCFA-producing microbes by providing extra microbiota-accessible carbohydrates—the main component of dietary fiber. Specifically, I suggested that he cut down on red meat and instead eat protein-, fiber-, and polyphenol-rich plant foods—lentils, beans, grains, and a variety of fruits and vegetables.

I told David that regular, moderate exercise has a favorable impact on gut health and an anti-inflammatory effect on the gut's immune system. The words *regular* and *moderate* here are key: if your exercise is sporadic, it may not be worth the effort, but if you exercise too strenuously and your gut is vulnerable to such physical stress—as David's was—you may reverse healthy results. As with a plant-based diet, the benefits of this type of exercise routine are largely mediated by an increase in gut microbial taxa that increase SCFA production, which strengthens the integrity of the gut wall and reduces gut-associated immune activation.

David's questions led me to further investigate the relationship between what we eat, how we exercise, and how the two might support each other. Given that both diet and exercise result in similarly positive adaptations of the gut microbiome and its communication with the body and the brain, I wondered if it was possible that, contrary to the prevailing sports dogma, a largely plant-based diet could trump a high-protein, animal-based diet for athletic performance.

The 2018 documentary *The Game Changers* tells the story of James

Wilks, elite Special Forces trainer and winner of the mixed-martial-arts competition reality show *The Ultimate Fighter*, as he searches the globe for the best athletic-performance diet. After consulting with top athletes, special-ops soldiers, and visionary scientists, Wilks ultimately concludes that, in contrast to the deeply ingrained belief that eating large quantities of animal proteins is essential to athletic achievement, a plant-based diet not only provides the same amount of protein, but may also be superior for optimal performance. This point wasn't lost on the ancient Romans, who fed their gladiators and soldiers a mostly vegetarian diet. Although much of the evidence proffered in this popular and highly influential film is based on anecdotal information as opposed to established science, many athletes I know have consequently changed their eating habits, and none has experienced any decline in performance.

Embriette Hyde, a science writer and former project manager for the American Gut Project at the University of California in San Diego, an avid athlete herself, created a small, uncontrolled study that more scientifically affirms this theory.[9] She decided to assess whether the dietary habits and gut microbiomes of several elite athletes could elucidate the role that the gut microbial ecosystem might play in these athletes' unique performances.

First, she evaluated stool samples from a group of extreme athletes, including climbers and mountaineers Alex Honnold, Emily Harrington, and Adrian Ballinger; runners Rob Carr and Amelia Boone; skier Cody Townsend; and surfer Fergal Smith. Then she compared these athletes' relative abundances of microbial genera with a database of fifteen thousand stool samples, many from the American Gut Project. Several of the athletes ate a largely plant-based diet, while others ate variable amounts of meat. The majority had an elevated relative abundance of microbial genera that break down fiber into short-chain fatty acids. Alex Honnold was a standout, with an overwhelming abundance of *Prevotella*, reflecting his largely plant-based diet. Could it be that the large-scale production of short-chain fatty acids by the microbes in Honnold's gut played a role in powering his superhuman ascents of sheer rock faces, like his dramatic record-breaking climb of El Capitan, as captured in the award-winning documentary *Free Solo*? Emerging scientific research indeed suggests that his microbes may be lending a helping hand.

A recent study led by Jonathan Scheiman of the Department of Genetics at Harvard Medical School offers a partial answer to this question.[10] By studying the gut microbiome in runners before and after they'd run the Boston Marathon, these researchers found a high abundance of the genus *Veillonella* in some but not all of the athletes compared to a control group of sedentary subjects, and an increase in these athletes after their run. Obviously, unlike David, these runners clearly weren't vulnerable to the detrimental effects of endurance exercise on the gut. Not everyone is as susceptible, and various types of exercise can have different effects on the body. When the researchers isolated a strain of *Veillonella* from the runners' stool samples and introduced it into mice, this transfer significantly increased the mice's treadmill running time, suggesting that a metabolite produced by *Veillonella* during extreme exercise might be responsible.

Lactate is a substance formed by muscle tissue as the body breaks down carbohydrates for energy, especially during intense exercise, and it turns out that the gut microbe *Veillonella* utilizes it as its sole energy source. When investigators performed an analysis of the elite athletes' genomes, they found that every gene in a major pathway that metabolizes lactate into the SCFA propionate showed increased expression after exercise. This propionate is released into the gut, where it is absorbed into the bloodstream.

The scientists also showed that the exercise-induced lactate from the blood can leak into the lumen of the gut and come into contact with certain microbes, including *Veillonella*. When the researchers transplanted fecal material with increased SCFA from the athletes into the mice, the rodents again logged more treadmill time. More research is needed to determine the mechanism by which the increased SCFA improves athletic performance, but it is likely an additional energy source for our muscles. These studies reveal that a particular gut microbial strain improved the mice's treadmill performance by converting the extra lactate induced by exercise into a new energy source.

Through their intriguing studies, these researchers identified a natural, microbiome-related chemical transformation in the gut that enhances athletic performance. Not only is the genus *Veillonella* enriched in athletes after exercise, but the pathway this microbe uses for the conversion of lactate is also enriched.

A chicken-or-egg question remains: Does the native microbiome of marathoners make them better athletes, or does marathon training change the composition of their microbiomes in a beneficial way? Scheiman's group proposed that the high-lactate environment produced by athleticism offers the advantage of either creating more lactate-metabolizing organisms, such as *Veillonella*, or increasing the microbes' metabolizing capacity, or both, resulting in greater endurance. Perhaps in some of the athletes diet influences the microbiome in a similar way. In any case, it seems evident that a plant-based diet rich in microbiota-accessible carbohydrates (MACs) with minimal amounts of easily absorbable sugar leads to increased gut microbial SCFA production, which not only contributes to greater microbiome health, but also, in people with high *Veillonella* levels, yields an added shot of energy during intense exercise. This diet can aid exercisers and athletes of all levels.

## How Food and Mood Affect the Microbiome

We all face challenges that can trigger the brain's stress-response system, which can influence the gut and its microbiome. We also encounter smaller stumbling blocks that don't raise alarms in most of us but engage the fight-or-flight response in people with "increased stress perception," creating a cascade of issues beyond the obvious. Research has consistently shown that people's thoughts and feelings about health affect both behavior and outcome. Our mind-sets are so powerful that they can shape a whole range of consequences—from the effects of exercise to the impacts of stress and diet to the lengths of our lives.

Studies have shown that people who think aging inevitably leads to physical or mental deterioration actually die sooner than people with a more positive attitude. One investigation found that those who don't view stress as harmful were the least likely to die compared to other groups in the study—including those who actually experienced very little stress.[11] Dr. Alia Crum, an assistant professor of psychology at Stanford University and head of the Stanford Mind & Body Lab, led a study that found that finance workers, observed during the height of the 2008 financial crisis, who believed that stress enhanced their ability to work experienced healthier physiological responses to demands than their counterparts operating under the assumption that stress is debilitating.[12] This same research group also reported that hotel-room attendants who adopted the view that their work is good exercise later showed greater reductions in weight and blood pressure than attendants who didn't regard their work in this way. In these situations, the brain's interpretation of the circumstances and the workers' attitude toward the work had more influence on their well-being than the work itself.

Likewise, people's perceptions about the positive and negative consequences of eating certain foods can influence their reactions to those foods. In our current age of abundant misinformation, with sham "diet science" littering websites and streaming through social media feeds, perceptions of particular foods as dangerous can spread far and wide. Michael Pollan referred to this burgeoning trend almost two decades ago as "our national eating disorder."[13] The emotional component of our food choices is expressed in an array of behaviors. Some are considered clear psychiatric disorders, such as orthorexia (an unrealistic obsession with finding the perfect health foods), anorexia, bulimia, and food-related phobias. Others are the "distinctly American" food fads cited in Pollan's article, including lipophobia (fear of fat), carbophobia (fear of carbohydrates), and many self-diagnosed but unproven food sensitivities.

All of these problems, including the psychiatric ones, have in common the prominent risk factor of "trait anxiety"—the consistent perception of the environment as threatening. This is present in a given individual from an early age, and it increases the risk of developing a wide range of other psychiatric disorders. In those who develop neuroses about diet, this abnormal underlying anxiety and hyperresponsiveness triggers a stress response, with all of its problems for the gut and its microbiome. In his article, Pollan writes about Paul Rozin, a University of Pennsylvania psychologist who, with French sociologist Claude Fischler, has extensively studied cross-cultural differences in attitudes toward food. Rozin and Fischler suggest that our distorted mind-sets and reflexively anxious eating are a distinctly American problem, especially among those at the high end of the socioeconomic spectrum. A more relaxed, social approach toward eating, as is the norm in many other cultures, could go a long way toward breaking our unhealthy habits of gorging and fad-dieting.[14] Of the four populations Rozin and Fischler surveyed—American, French, Flemish Belgian, and Japanese—Americans derived the least pleasure from eating. This is because eating in a pleasant social setting as a festive occasion amongst family or friends, rather than in the car or in front of the television, makes people feel good. Neural pathways in the brain give us hedonic pleasure when we enjoy delicious food in pleasant company without worry or guilt feelings.

## A Cross-Cultural Eating Cure

In my practice, I've seen many patients with chronic digestive symptoms. I vividly remember Kristen, a pleasant young woman in her early twenties, who came to my office with her father to treat her distressing symptoms of abdominal bloating and constipation. Kristen was a senior at an Ivy League college with a double major in business and Italian. She was planning to apply to law school in the fall. In addition to having a heavy course load, Kristen was also on the varsity swimming team. Even though she'd suffered from bouts of anxiety during her senior year of high school, she hadn't experienced digestive symptoms until midway through her freshman year of college, just as she'd begun her vigorous exercise program with the swim team. Kristen was primarily concerned about bloating, which caused a visible distension of her abdomen and occasional bouts of nausea. She was self-conscious and embarrassed about the way she looked when this happened, especially at swimming practice and when she was socializing.

Prior to coming to my office, she'd seen several doctors and dietitians who recommended different treatments—including a gluten-free diet and a low-FODMAP diet, which reduces beneficial fermentable fibers found in beans and legumes but is popular with many doctors for its reduction of gas, bloating, and irritable-bowel syndrome. For Kristen, though, no diet provided much relief from her symptoms.

As we chatted, Kristen mentioned that she'd recently returned from a semester abroad in Florence, Italy. "I loved Florence. It was probably the best time of my college years so far. And, amazingly, all of my digestive problems practically disappeared while I was there. At first I was afraid to eat gluten-containing foods but then it seemed crazy not to eat pasta in Italy, so I gave in and not only ate pasta but also bread and pizza and all types of vegetables—and without any bloating. It was wild. After a few weeks of that, I completely lost my fear of gluten!"

I told Kristen that she wasn't the first patient who'd experienced such a surprising shift in digestive symptoms while traveling. "What happened when you got back to the US?" I asked.

"I was already worried on the plane back to LA," she confessed. "I was scared that my symptoms might come back—that somehow it was only in Italy that I could magically eat like that without any repercussions." On the flight home, she ate vegetarian lasagna, along with a roll and a small dessert. Obviously, the lasagna didn't come close to the delicious pasta she'd enjoyed almost daily in Florence, a disparity that only ramped up her worry that her recovery had been too good to be true.

Lo and behold, a couple of weeks after her return, all of her old symptoms returned, and she began to have the same obsessive anxieties about what might happen when she ate. Everything seemed to make her feel bloated and distended. "Looks like they fed you well in Italy!" someone on her swim team joked one day at practice, a remark that was not only humiliating also but confirmed for Kristen that her problems were as noticeable to others as they were to her.

I explained to Kristen the powerful influence our brain can exert over the gut and its microbes. I told her it was clear, even in our brief conversation, that she was under tremendous stress, as she was not only excelling academically but also pushing herself to her physical limits. By contrast, it sounded as though she'd enjoyed a much more relaxed lifestyle in Italy, with enough free time to read a book or hang out in a café with a cappuccino or gelato with friends, things she didn't allow herself to do at home. High-quality Italian food may have played a role in her temporary digestive health, but the fact that her chronic symptoms didn't depend on any particular food made it much more plausible that food itself wasn't the main culprit.

I told Kristen I suspected that her chronic stress and food-related fears had altered the interactions within her brain-gut-microbiome network. I described the research about extreme physical exercise and its negative effects on the gut, especially when combined with stress, and I recommended that Kristen set up a consultation with our wellness coach to discuss the possibility of a short course of cognitive behavioral therapy. I hoped this would help her to diminish her food-related anxiety and decrease the pressure she'd been putting on herself. Then I advised her to relax her extreme exercise regimen. In her case, that meant swimming only in school but not doing additional training in a private club.

Experiences with food are rarely objective. Kristen's story highlights the emotions involved when one is chronically obsessed with gauging the health effects of certain foods. Eating becomes a chore instead of a pleasure. Numerous studies have illuminated the psychological influences that may have played a role in Kristen's experience, demonstrating that the exact same foods can be experienced as tasty, filling, and rewarding or bland, skimpy, and repulsive, depending on how they're described to people before they eat them. Research has also shown that perception *can* lead people to eat healthier foods. In 2016, Kaitlin Woolley and Ayelet Fishbach, researchers from the Booth School of Business at the University of Chicago, published a study in which some participants were prompted to "choose the carrots you think are the tastiest and that you will enjoy eating the most." They consumed more than those encouraged to "choose the carrots you think are the healthiest and that you will benefit most from eating."[15] Likewise, in another study from the same group, young children consumed more carrots when told a story about a character who had experienced their delicious taste as opposed to one about achieving a particular goal by eating them.

Stanford's Dr. Alia Crum—who has researched how our attitudes toward stress determine our response to it—also led one of the largest and most comprehensive studies evaluating how much a positive mind-set about eating can influence our selection and consumption of healthy foods. In collaboration with the Menus of Change University Research Collaborative (MCURC), Dr. Crum's lab performed the DISH (Delicious Impressions Support Healthy eating) study, a randomized, controlled intervention study conducted in five university cafeterias throughout the United States. Researchers tested whether more indulgent, taste-focused labels (like SWEET SIZZLIN' GREEN BEANS AND CRISPY SHALLOTS) would influence the amount of vegetables people ate as compared with more health-focused descriptions (like LIGHT 'N' LOW-CARB GREEN BEANS).[16] In 137,842 diner decisions over 185 days and twenty-four vegetable types, the taste-focused labels increased vegetable selection by nearly one-third compared with health-focused labels and by 14 percent compared with basic labels (GREEN BEANS).

The researchers were also able to show that it was the higher taste expectations that made people gravitate toward the taste-focused labels. Tasty labels outperformed the merely positive and the ones with fancy words or even lists of ingredients. The authors concluded that manipulating attitudes about food by emphasizing its tastes, aromas, and textures can increase the amount of vegetables people eat, even when the veggies are competing with less healthy options that we're conditioned to find more appealing.[17]

We already know that anxiety, depression, and stress can disrupt the brain-gut-microbiome network, increase intestinal permeability, and change microbial composition and function in various ways. Though there hasn't yet been a study illustrating a direct effect of mind-set on gut microbial composition and function, I have no doubt that the attitudes we hold about food—and the associated stress and anxiety—can have an important influence on the gut, creating a mirror image of this attitude within the gut and its microbiome.

Ultimately, how we feel affects what we eat and what we eat affects how we feel. Our diets *and* our mind-sets have a major influence on the gut microbiome. What we eat determines which microbes will benefit, and that metabolic choice is communicated throughout the body and brain via the BGM network. Food-related fears and the stress of choosing the *right* food—not to mention stress and anxiety in general—can modify microbial composition and function via signals to the gut microbiome via the autonomic nervous system. In addition to many other unknown changes, such changes include a reduction in the abundance of *Lactobacillus* and in the overall diversity of the microbiota.

## Sleep

Despite the well-established correlation of physical and mental health with sufficient sleep, as a culture we've made sleep deprivation an accepted price of modern life. From the reported 73 percent of high-school students who don't get enough sleep to the shift workers who must stay awake on the job despite unnatural hours to the go-getters proudly claiming they need only a few hours of sleep a night, we are a culture in need of an awakening about the importance of sleep. In 2017, some 35 percent of Americans reported their sleep quality was poor, despite the fact that, like bad meals and scant exercise, poor sleep comes with increased stress and irritability, as well as greater risk of metabolic syndrome, cardiovascular disease, cancer, and infections. Sleep plays a crucial role in regulating the immune system and exerts a systemic anti-inflammatory influence. As the gut contains about 70 percent of the body's immune cells and is closely connected with the brain via neural and chemical pathways, one would expect that the BGM network would play an important role in the regulation of sleep.

Indeed, sleep is essential to healthy gut function. When we aren't actively eating or digesting food, our gut microbes are forced to switch temporarily to another fuel source—in particular, the complex sugar molecules, or glycans, that make up the gut's mucus layer. Although a chronic reduction of the thickness of the mucus layer in response to an unhealthy diet leads to a leaky gut, oscillations in this barrier between day and night are also part of a healthy gut physiology, allowing intermittent communication between the gut microbiota, the gut itself, and other organs.

When we are at rest, the gut switches from its regular peristaltic pattern of back-and-forth contractions to a cyclical, high-pressure propulsive pattern called the migrating motor complex. During this time, a band of high-amplitude contractions originating in the esophagus slowly makes its way down to the end of the small intestine, taking with it undigested food particles, intestinal fluids, and trillions of gut microbes, which are then swept into the large intestine. One of the many functions of this motor wave, which in the fasting state recurs every ninety minutes, is to keep the density of microbes in the proximal small intestine (the part closest to the stomach) low, while leaving the density in the large intestine untouched, preventing SIBO, or small-intestine bacterial overgrowth.

As far back as 350 BC, Aristotle observed in his book *On Sleep and Sleeplessness* that sleep is induced by influences originating from the stomach during digestion and that it also can be triggered by high body temperature. Even though Aristotle had no scientific knowledge of the intricate interactions between the immune system, inflammation, and the brain mechanisms underlying sleep, he described a slumber-inducing response in feverish patients—the first prescientific description of the sleep-immune interaction. Sleep-immune interactions are well-known phenomena in everyday life and in folk wisdom. We've all experienced exhausting illness followed by a good night's sleep. After all, it's "the best medicine."

In the early twentieth century, researchers postulated that a molecule called hypnotoxin increases during wakefulness, induces sleep, and is then cleared during sleep. It was later discovered that this presumed sleep-inducing molecule was a component of a bacterial cell wall (a lipopolysaccharide, or LPS) and was thought to originate in the gastrointestinal tract—making Aristotle the earliest believer in the brain-gut-microbiome network. By activating the immune system and releasing sleep-regulatory substances, including the "immune system soldiers" known as cytokines, these microbial cell-wall components were shown in animal models to contribute to the homeostatic regulation of slow-wave sleep,[18] our deepest form of sleep.*

As I explained in chapter 1, an increase in cytokine and LPS blood levels not only occurs during an infection, but has also been observed in metabolic endotoxemia, the noninfectious, low-grade immune activation that develops in response to an unhealthy diet and the resulting compromised gut barrier function. With our advanced understanding of the interactions between the gut microbiome and gut immune system, as well as the circular interaction of these systems, a more complete perspective on the relationship between poor sleep, the gut microbiome, and chronic illness has emerged.

Christoph Thaiss, assistant professor of microbiology at the Penn Institute of Immunology, Perelman School of Medicine, University of Pennsylvania, reported results performed during his postdoctoral work with Eran Elinav at the Weizmann Institute in Tel Aviv that showed the timing of *when* a person (or mouse) eats in relation to circadian (day and night) rhythm plays a critical role in shaping intestinal microbial ecology and gut health.[19] When food intake was rhythmic—that is, when subjects ate corresponding to their circadian rhythms (which means eating during the daytime for humans and nighttime for mice)—researchers found that in the same individual

---

* During sleep, the brain moves through five different stages.

One of these stages is **rapid-eye-movement (REM) sleep**. The other four phases are referred to as non-REM sleep. REM sleep is characterized by rapid movement of the eyes, fast and irregular breathing, increased heart rate and blood pressure, increased oxygen consumption of the brain, brain activity similar to that seen while awake, and sexual arousal in both men and women. REM-sleep alterations can be one of the earliest symptoms of Parkinson's disease.

**Slow-wave sleep** is the deepest phase of non-REM sleep and is characterized by particular kind of EEG activity (**delta waves**). During slow-wave sleep, dreaming and sleepwalking can occur, and this sleep phase is thought to be important for memory consolidation.

around 15 percent of all types of microbes living in the gut fluctuated in abundance over the course of a day, while the other 85 percent remained relatively constant in their numbers. These variations of microbiota populations between day and night, which are similar to oscillatory biological mechanisms in the liver and gut, are influenced by the circadian clock in the suprachiasmic nucleus of the brain, as discussed in chapter 3. These variations are accompanied by changes in the way the gut microbiota interact with gene-expression patterns in the gut cells; they perform critical functions in adapting metabolic processes throughout the body to the day-night cycle.

The investigators showed that the disruption of the normal circadian rhythm in mice leads to a state of dysbiosis, an aberrant microbiota composition. To determine if the sleep-wake cycle similarly affects the human gut microbiota, these researchers studied jet lag in people flying between countries with an eight-to-ten-hour time difference. They collected fecal samples from two healthy human donors willing to undergo this flight-induced time shift and studied their relative microbial abundances one day before the start of their travel, one day after landing, and after their recovery (two weeks after landing).

As the investigators had hypothesized, the microbiota communities of the two travelers showed a jet lag–induced change in composition, with a higher relative abundance of the phylum Firmicutes, which has been associated with a higher risk of obesity and metabolic disease in multiple human studies. However, this shift was promptly reversed upon the participants' recovery from jet lag. In an effort to determine whether these microbiota changes could lead to increased susceptibility to metabolic disease, the investigators then transferred the fecal samples into germ-free mice. The mice subsequently experienced weight gain and higher blood-glucose levels as detected by an oral glucose-tolerance test, which measures the body's response to ingested sugar and is sometimes used as a screen for type 2 diabetes. This metabolic disturbance was reversed by the transfer of fecal material from the human participants after jet-lag recovery.

Given these findings—along with the body of research proving the detrimental effects of inadequate sleep on the immune system, including an increased risk of a host of illnesses—it's clear that getting a good night's rest is as important to a normal functioning of the BGM system and to our long-term health as exercising, eating right, and adopting a positive attitude toward food. In the next chapter, I'll further explore the latest research about the profound relationship between the gut and the internal clock, showing that *when* to eat is as important as *what* to eat.

**Chapter Seven**

# RESTORING THE GUT MICROBIOME

The dramatic shift in our Western diet, starting with the accelerated rise of industrial agriculture after World War II and culminating in round-the-clock access to cheap, ultraprocessed food packed with animal fats, refined sugar, and nutritionally empty calories—along with our increasingly sedentary lifestyles—has played a central role in the advance of many of our chronic diseases. There is compelling research that specifically puts disturbances to the gut at the center of this health crisis. The good news is that we can reverse this devastating trend by adjusting what and when we eat.

Even for the extremely health conscious among us, there's a new way of viewing *how* we eat that current diet trends haven't yet fully incorporated. Many of us are still focused on macro- and micronutrients—carbohydrates, proteins, fats, and vitamins and minerals. I encounter this focus professionally in my work with patients, of course, but also personally. In fact, I find myself in a recurring conversation about nutrition with my hiking and running partner, Rich. In his twenties, Rich was a member of the US Olympic men's gymnastics team and has been an active rock climber since his youth. Now in his early seventies, he's still amazingly active, with an athletic build that many forty-year-olds would envy. Not too long ago, he became a vegetarian. In fact, he made this decision after watching *Game Changers*. He tells me that the shift in his diet has not negatively affected his physical strength or athletic ability. However, as a late-blooming convert to a largely plant-based diet, Rich's food choices are now largely based on concerns about getting enough protein.

As I often tell him, as well as patients with similar worries, eating adequate protein isn't a problem for most people, even vegetarians. Around the world, on average, humans consume about 30 percent more protein than the officially recommended daily amount, which is 0.36 grams of protein per pound of body weight per day, or about two ounces of pure protein for a 150-pound person, bearing in mind that protein content varies widely and no natural food is pure protein. In fact, North Americans and Europeans consume about twice that amount.[1]

The recommended daily amount is based on the "zero nitrogen balance" concept, meaning the amount of protein ingested should provide the amount of nitrogen naturally lost through urine, skin, and hair. For most of us living in the developed world—including my friend Rich—there's no need to worry about getting enough protein or to spend money on high-protein energy bars, shakes, and supplements. Unfortunately, the situation is vastly different in the developing world, where undernutrition and insufficient protein intake *are* real problems with serious health consequences.

It's interesting that it's not the total amount of protein that matters, but rather the source. A recent study led by Drs. Jiaqi Huang and Demetrius Albanes, from the National Cancer Institute, followed participants for sixteen years and showed reduced mortality for those who ate a higher percentage of their daily protein (an average of fifteen grams per day) from plant sources as opposed to animal sources.[2] The study included 237,036 men and 179,068 women with a median age of sixty-two. Higher plant-protein intake was associated with reduced risk of overall mortality, about 12 to 14 percent lower for each ten-gram or thousand-calorie intake increment. The benefit was apparent for cardiovascular disease and stroke mortality in both sexes and was independent of several risk factors. Replacement of 3 percent of energy from various animal protein sources with plant protein was associated with 10 percent decreased overall mortality in both sexes. Among the various protein sources, replacement of egg protein and red-meat protein with plant protein resulted in the most pronounced reduction in mortality—24 percent for men and 21 percent for women.

But reduced mortality isn't the only reason I tell Rich he ought to reorient his diet attentions away from the concern about not getting enough protein. A revolutionary new perspective has developed about exactly *whom* we're feeding when we eat.

Consider your microbes. While macro- and micronutrients—fats, proteins, carbohydrates, vitamins, and minerals—*are* essential, in the vast majority of healthy people, they're all rapidly and effectively absorbed in the small intestine. They never make it farther downstream to the gut microbes, living in the dark, oxygen-free environment of our large intestine. Though the number of microbes increases the farther down you move in the small intestine, from the duodenum to the jejunum to the ileum, the greatest population lives in the large intestine.

Until recently, most nutrition research has focused on nutrient absorption in the small intestine, while the gut microbiota in the large intestine have been largely ignored. That's why we haven't been advised till recently to eat more foods of low caloric density containing nondigestible components, such as the fiber found in most fruits and vegetables. This fiber can't be broken down by our own enzymes in the small intestine and therefore can't be rapidly absorbed, allowing it to travel all the way to the most heavily populated and microbially diverse areas in our large intestine. This new eating strategy—prioritizing your microbes—not only benefits the health of the gut and microbiome but also, as network science tells us, supports the healthy function of *all* organs, including the brain.

This nonabsorbable part of our diet is crucial to our health from the moment we enter the world. Many experts recommend that new mothers breastfeed because of the beneficial fats and calories in breast milk, but it's actually the nonabsorbable portion of the milk that provides the greatest benefit to an infant's developing brain-gut-microbiome network.[3] Certain complex carbohydrates called human milk oligosaccharides (HMOs) are too big to be absorbed in the small intestine. Babies don't even have the enzyme necessary to digest oligosaccharides because they're intended solely for the developing gut microbial ecosystem in the baby's colon.[4] There they play a critical role in coordinating the assembly of a healthy microbiome. As J. Bruce German, the late food scientist from the University of California at Davis, once remarked to Michael Pollan, "Mother's milk . . . is telling us that when natural selection creates a food, it is concerned not just with feeding the child but the child's gut bugs too."[5]

What is the optimal diet to feed our gut bugs? While nutrition science is constantly evolving, a meta-analysis recently published in the *British Medical Journal* offers useful insight. The study authors compared the results of fourteen popular diet programs for both weight loss and reduction of cardiovascular risk, dividing them into three categories: Low Fat (such as the Ornish diet), Low Carb (Atkins, South Beach, The Zone), and Moderate Macronutrients (including Biggest Loser, Jenny Craig, Weight Watchers, and Mediterranean). They then examined the results at six months and twelve months.[6] Most of the diets offered substantial improvements in cardiovascular risk factors, particularly blood pressure, as well as modest weight loss at the six-month mark. By the twelve-month follow-up, these effects had largely disappeared for all popular programs *except* the Mediterranean diet. Most positive diet effects fade over time due to diminishing compliance, but this wasn't the case for the Mediterranean diet. Only this largely plant-based diet, which is both satisfying to humans and nurturing to the microbes within us, showed a statistically significant difference at twelve months in weight loss and improvement in cardiovascular risk factors, including a reduction in low-density lipoprotein (LDL) cholesterol, the "bad" cholesterol.

Instead of worrying about protein, carbs, and fat, it's clear that we would be better served by focusing on the long-neglected needs of our invisible microbial population. For adults, this means consuming non-absorbable dietary fiber, polyphenols, plant-derived anti-inflammatory food components, and other large molecules that are components of a plant-based diet. All of these can *only* be metabolized, broken down into smaller molecules, by the biochemical machinery operated by various gut microbes in the end of our small intestine and especially in the large intestine. The process yields hundreds of thousands of metabolites, which beneficially support every part of the BGM network, acting directly on the nerve, immune, and endocrine cells of the gut, as well as on vagal nerve pathways between gut and brain. These molecules can act locally on the gut or be absorbed to reach the brain and other organs through the bloodstream. Many of these metabolites have been identified, and intense research efforts are under way to build huge databases of biochemical information about these molecules in hopes of identifying new disease mechanisms and treatments for obesity, depression, Alzheimer's, Parkinson's, and other diseases.

## The Dual Evolution of Our Diet and the Brain-Gut-Microbiome Network

Just as our current levels and types of stress have made our age-old fight-or-flight response system maladaptive during the last seventy-five years, a similar problem has also occurred in the evolution of our BGM network. Our eating habits and the BGM network have co-evolved. Until recently, dietary changes have occurred slowly enough to give the human digestive tract and brain enough time to adjust their structure and function. "Enough time" for these mutual genetic adaptations is about ten thousand to thirty thousand years. They have occurred slowly but surely over the course of our evolution, from the diet of the earliest hominid hunter-gatherers of millions of years ago, to the advent of using fire to cook food several hundred thousand years ago, to the agrarian revolution that began the Neolithic (New Stone) Age twelve thousand years ago, when we transitioned from foraging to farming. The evolution of our bodies and our bacteria continued more or less in synchronicity until industrialization came along in the nineteenth century, introducing the first wave of processed foods.[7]

At least two million years ago, the hunter-gatherer culture developed among the early hominins of Africa. Like their nonhuman primate cousins, these early ancestors had small brains (requiring less energy) and large intestines packed with trillions of microbes, allowing them to break down and absorb large amounts of otherwise indigestible food components. Their digestive system was optimally adapted to their environment, extracting energy from fiber molecules of plants as well as from the rest of the plant food and from meat. These omnivorous hunter-gatherers also ate a lot of animal protein; they killed animals instead of scavenging meat left behind by other predators. But they also ate various grasses, tubers, fruits, seeds, and nuts. In fact, the examination of a human settlement site in Israel from almost eight hundred thousand years ago revealed remnants of *fifty-five* different food plants, along with evidence that fish was a part of the diet.[8]

Sometime between eight hundred thousand and three hundred thousand years ago, humans began cooking their food. This caused such a revolution in the way we ate that John B. Furness, a professor at the University of Melbourne and a pioneer in studying the enteric nervous system and the gut connectome, coined the term *cucinivore* to distinguish the new dietary habits of this era from those of the previous omnivores.[9] A few of these pre-Neolithic hunter-gatherer societies actually still exist in the world today, including the San (formerly called Bushmen) of southern Africa, the Sentinelese of the Andaman Islands in the Bay of Bengal, the Hadza of East Africa, and the Yanomami of the upper Orinoco River.

Cooking not only changed our form and style of eating, which altered social behaviors, bringing people together around the magical hearth, but it also introduced dramatic changes to the structure and function of the BGM axis. Although "processed foods," in particular those containing emulsifiers, fructose, artificial sweeteners, and added gluten, are considered modern-day villains, the original meaning of this term actually refers to the period when humans began to cook and store food. The original processed food was one of humankind's most revolutionary inventions. Before it, hunter-gatherers had to graze almost continuously in order to consume enough calories to function. However, with the advent of cooking, humans were able to take in more calories more quickly, as food was now easier to chew, digest, and absorb in the small intestine. Increased energy intake led to the brain evolving faster, while the large intestine, no longer essential to ferment large amounts of unprocessed food and turning it into absorbable calories, became smaller.

Once humans were able to process and preserve foods, a number of major nutrient sources became more readily available. The most obvious example is grains—which we started cultivating and growing, making it a common food source, at least ten thousand years ago.

Even though humans were foraging naturally growing precursors of ancient grains at least forty-five thousand years ago, they weren't viable food sources to support early humans' increasing caloric needs, as all mammals lack the enzymes necessary to break down unprocessed grain starches. Many people haven't heard of einkorn wheat, dinkel wheat (spelt), bulgur, farro, emmer wheat, buckwheat, or khorasan wheat (now trademarked as Kamut) and most rarely eat dishes made from millet or sorghum. Compared to modern grains, all of these ancient grains have more large, complex sugar molecules—the microbiota-accessible carbohydrates (MACs) that feed our gut microbes—and fewer simple carbohydrates, which are easily absorbed in the small intestine.[10] In order to store foods for longer periods of time, humans learned to control fermentation, an anaerobic (oxygenless) process in which various benign microorganisms live in the food and prevent the growth of decay bacteria. This early method of preservation happened to provide unanticipated health benefits as well, such as enhancing the natural, beneficial bacteria in the food and when ingested regularly in sufficient amounts adding to the diversity and richness of the gut microbiota. In fact, the resulting adaptations of our gut microbiome to naturally fermented foods offers a strong argument in support of the consumption of naturally fermented foods today.

Most animals are confined to a narrow range of diet based on genetically determined physiology. Cows must eat low-protein fibrous plants. Cats are obligate carnivores, meaning that they can't digest anything but meat (although they do chew grass, possibly as a gut cleanser or for folic acid). Koala bears eat only eucalyptus leaves. Humans, on the other hand, can eat a wide variety of energy sources. By expanding our food repertoire through preservation and cooking, we not only diversified what we eat, but also induced more digestive flexibility. Perhaps most important, the communal time around the hearth cooking, eating, and talking fostered increased social interaction, cooperation, and brain development—which Kristen still cited from her gut-healing meals in Florence as a benefit of the Mediterranean diet (see page 113).

However, over the last 75 years or so, we've developed other processing methods, which have packed our diet with *ultra*processed food, loaded with an unprecedented amount of sugar (including new forms, such as high-fructose corn syrup), preservatives, artificial flavors, emulsifiers, and added gluten, just to name a few. These changes have occurred along with new methods of cooking and preservation, such as heat sterilization, microwaving, refrigeration, and irradiation, all of which can affect the abundance of food-associated microbes.

For example, consider the ancient grains, which introduced beneficial MACs into our diets, a veritable feast for our microbes. But now, just yesterday in evolution time, we began to genetically select and highly process these grains, making them widely and commercially available but dramatically reducing their diversity while removing most of their indigestible fiber and reducing their micronutrient content. These ultraprocessed grains, which would have been unrecognizable to our ancestors, now make up an estimated 70 percent of the dietary intake in modern societies. In today's average American diet, an estimated 78 percent of all food energy intake, or about 1,000 calories per day, comes from moderately processed or ultraprocessed foods derived from a smaller and smaller number of plant species.

This abrupt change in our diets combined with the gradual, structural shifts in our BGM network—our shorter GI tracts and colons—has created the current damaging mismatch between what we eat and how our bodies respond to it. Today, the total size of the human GI tract in relation to body weight is about half of its size in other mammalian species.[11] More important, the colon represents only 20 percent of the total volume of the digestive tract, whereas in our primate cousins it was about 50 percent, allowing for substantially higher energy extraction, by fermentation of plant fibers and their transformation into absorbable short-chain fatty acids.[12] In fact, today's human "hindgut"[13]—the end of the small and the entire large intestine—accounts for only 6 to 9 percent of energy extraction, compared to up to 50 percent in other mammals, such as horses.

Ultimately, this change has resulted in the stunning fact that the trillions of gut microbes in the human colon have been rendered inessential for deriving energy from food. This dramatic shift—making the proximal small intestine the major site for energy harvest in our bodies—has had major consequences on the composition and richness of our gut microbial ecosystem, which rapidly adapted to the change in dietary habits. The changes of our gut and its microbiome were clearly an adaptive evolutionary development for humans in response to the invention of cooking. With the introduction of ultraprocessed food, we've pushed this adaptation past its limits, and we're now paying the price. With our relative recent dietary changes we have reached a point where what was once good for our ancestors is no longer good for us.

The recent shift in this network dynamic is like the changes caused by a large dam to the ecosystems connected to the river. Before the construction of the Hoover Dam, the water, fish, and energy of the Colorado River was shared by people and businesses all along it from the Rocky Mountains in Colorado to its delta in northwestern Mexico. The US government built the dam to generate hydropower and supply electricity to the Western states. However, the reduced flow downstream has since decreased Colorado Valley farm output in Arizona, California, and Mexico, impoverishing the villages dependent on it. In an analogous way, most of our ultraprocessed diet is absorbed in the small intestine, and the trickle of undigested food components that makes its way to the "delta" of our large intestine can't feed many microbes. The whole ecosystem suffers. Our microbes are forced to turn to a different source for large molecules—such as the sugar molecules making up the mucus layer of our gut—eroding the barrier and creating disruptions throughout the entire BGM network.

While the discovery of early food processing and resulting gradual expansion of our dietary options played an important role in our biological and cultural evolution, the unprecedented and dramatic changes in our food supply resulting from the accelerated industrialization of agriculture have inadvertently led us toward the health-care crisis we find ourselves in today. The question I am frequently asked, is if there's also a map to chart our course back to health.

## What to Eat

Based on a wealth of scientific and clinical evidence, I've devised a healthy food plan that offers easy, direct ways to choose what to eat so as to restore the balance of the gut and achieve greater health overall. The recommendations I make here and in chapter 10 are based on what I call the healthy-food index, the proportion of microbiome-targeted foods in the total diet. The more polyphenols, plant-based fiber, phytonutrients, and complex anti-inflammatory molecules that we eat in a largely plant-based diet, the lower the caloric density of that food and the higher the healthy-food index will be. Conveniently, a diet of high-index foods automatically provides enough vitamins and other micronutrients in addition to healthy plant-based proteins, oils, and fats.

Eating a three-scoop hot-fudge sundae will deliver a whopping 750 calories of immediately absorbed sugar (45 percent) and fat (49 percent) to the small intestine, causing an increase in sugar levels, an insulin spike in the blood, and no leftovers for the gut microbiota. On the other hand, a mixed salad with beans, nuts, and avocado (without any cheese or high-calorie dressing) will provide some absorbable nutrients, like vitamins, to our small intestine, but the great majority of the salad will be delivered to the end of the small and to the large intestine, where the microbiome machinery will break it down into absorbable health-promoting molecules. The healthy-food index of the sundae is nearly zero, but that of the salad is high. As long as we concentrate on the best diet for our microbes, we'll also be looking after our own best health. Treat your microbes well and they'll return the favor.

## Fiber

When I went to medical school in the seventies, we were taught that the benefit of dietary fiber for digestive disorders came primarily from its ability to provide bulk and retain water in the intestine, resulting in more rapid transit of waste and improving bowel movements. In the 1960s and '70s, Drs. Denis Burkitt and Hugh Trowell documented the significantly higher intake of dietary fiber by Africans relative to Westerners (60 to 140 grams per day versus about 20 grams), reinforcing our understanding of fiber as an expedient way to relieve constipation.[14] The authors reported that rural Africans passed stool that was up to five times greater by mass, had intestinal transit times that were more than twice as fast, and ate three to seven times more dietary fiber than their Western counterparts, all this without having an elevated BMI. In the absence of a good biological explanation for the health benefit of a high-fiber diet, they pointed out the lack of Western diseases like diabetes, heart disease, and colorectal cancer in Africans. Today it has become obvious that the benefits of fiber go far beyond regulating our bowel movements. In contrast to earlier beliefs, "dietary fiber" isn't a homogeneous group of plant materials and molecules, nor are they all fermentable by microbes.* It's perhaps more accurate to refer to this group of complex molecules targeted

* The term *dietary fiber* encompasses hundreds of different types of complex sugarlike molecules. Dietary fiber is divided into soluble and insoluble types. Soluble fibers—such as the fructo-oligosaccharides (FOS) in garlic, onion, chicory root, artichoke, and asparagus—are shorter sugarlike molecules that dissolve in water and are rapidly metabolized by microbes in the end of the small intestine and the first part of the colon. Insoluble fibers—such as the cellulose in green vegetables like kale, Brussels sprouts, green peas, and oat fiber—are partially fermented in the second half of the colon by gut microbes, where the transit time is slower and the bacterial density is much higher.

at the gut microbes as microbiota-accessible carbohydrates, which you'll remember are a beneficial component of ancient grains. (There are no MACs in a Big Mac!) MACs are complex carbohydrates found in fruit- and vegetable-rich diets, and also contained in the gut's mucus layer. They're resistant to enzymatic degradation and absorption in the first part of the small intestine and so become essential food for our microbiota.

It's interesting that a fiber that serves as a MAC for one person may not have the same function for another, depending on the composition of each individual's gut microbial ecosystem, or *enterotype*. One person's microbes may have the set of enzymes required to break down a certain fiber molecule, while another may lack this particular microbial strain.[15]

Likewise, what may have been a MAC for our prehistoric ancestors may no longer be one for people in developed countries who've lost the microbial strains that metabolize it. For example, lactose can be metabolized and absorbed universally by the small intestine of infants around the world, but becomes a MAC in the majority of adults, who naturally lose the ability to metabolize it over time. In other words, the "lactose intolerance" that's now understood as a common medical disorder to explain nonspecific bowel symptoms such as bloating and indigestion is actually a natural physiological change. Some populations—Inuits and some Northern Europeans—are exceptions to this rule, as they've always consumed large amounts of milk and therefore maintained the lactase production required to break down and absorb dietary lactose. Japanese people harbor microbiota in their intestine that metabolize a particular source of fiber from marine microbes that feed on red algae and seaweed, making them the only people for whom seaweed is a MAC.[16]

Given this unpredictability, my advice to patients, in terms of knowing whether particular types of fiber are beneficial or not and which ones may cause discomfort, is to eat a large variety of fruits and vegetables, assuming that, though not all of them will be metabolized effectively by their individual set of microbes, a large number of them will. Over time, it's even possible for a progressive partial return and subsequent increase in microbial strains in the gut's ecosystem on a varied diet of large amounts of different dietary fibers. In the future, another option will be to use gut microbiome tests to determine for each individual which microbial metabolic pathways are available to break down MACs into beneficial metabolites and which food components can't be adequately processed.

Despite the declining role of our gut microbes in harvesting calories from indigestible carbohydrates, they remain essential for our overall, systemwide health. As they rapidly adapt to our modern dietary habits, they inform gut microbial signaling in our immune-activation, metabolism, and brain functions. Much of the energy that fuels the function of the gut microbiota comes from plant-derived complex molecules, which are made up of a large variety of interconnected simple sugar molecules, or monosaccharides. The way this energy is extracted depends on the symbiotic interaction of multiple microorganisms within a complex gut microbial ecosystem. Various microbes, each equipped with a specialized collection of enzymes from among those collectively called glycoside hydrolases and polysaccharide lyases, can break down the chemical linkages between these sugar molecules into consumable oligosaccharides or monosaccharides. These are subsequently taken up by other microbes and turned into absorbable SCFAs, such as butyrate, acetate, and propionate. As in any ecosystem, there's intense microbial competition for metabolic access to the energy and carbon sequestered in these molecules. Chronically reduced or abolished access to this source of energy will lead to the gradual decline and even extinction of certain strains—the very scenario we've found ourselves in.

In their 2014 article in *Cell Metabolism*, Erica and Justin Sonnenburg illustrated the complexity and function of this microbial food web, consisting of thousands of different strains of microbes interacting synergistically to process and utilize every piece of the food that reaches them.[17] The crucial importance of an intact gut microbiome for human health becomes obvious when comparing this impressive number to the small number of fiber-degrading enzymes that our own gut cells produce. Remember that millions of microbial genes contain the blueprint for the production of these enzymes, which enable the microbes to produce thousands of useful metabolites from what would otherwise be waste. This internal ecosystem has become known only in the last few decades, but it's perhaps the most important aspect of our diet!

Fortunately, the list of MACs available from a variety of plant-based foods is long. It's important to have a highly diverse microbiome in order to pull as many health-promoting molecules as possible from these foods.[18] Eating artichokes, beets, broccoli, lentils, and onions, for instance, will provide high amounts of MACs called galacto-oligosaccharides (GOSes). These will increase the relative abundance of strains of *Bifidobacteria*. By contrast, if you regularly consume cashews, white beans, oats, and sweet potato, you'll ensure the delivery of (digestion-)resistant starch to your gut microbiota, requiring more *Ruminococcus* and *Bacteroides*. Asparagus, leek, banana, garlic, chicory, and artichoke all have a lot of fructans (large molecules built of many fructose molecules—including the prebiotic fiber inulin and the fructo-oligosaccharides, or FOS). Eating them will draw on the *Bacteroides* and *Faecalibacterium* species and strains for optimal processing. If you love to eat apples, apricots, cherries, oranges, and carrots, you're delivering pectin to your gut microbes, which is processed by strains of *Eubacterium* in addition to the microorganisms specializing in fructans.

However, if you want to nourish and maintain the most diverse gut microbiome, containing the largest number of strains from all of the above microbial taxa, you'll need to eat a wide variety of *all* plant foods. Taking the popular and highly advertised shortcut of popping a daily supplement pill containing billions of colony-forming units (CFUs) will *not* do the job!

## Polyphenols

MACs aren't the only gut-health-promoting elements of diets rich in fruits and vegetables. In addition to micronutrients such as vitamins and minerals, and dietary fiber molecules, there are a variety of large plant molecules, collectively called polyphenols, that include families of compounds with exotic-sounding names such as flavonoids, anthocyanins, ellagitannins, and quercetin. Although structurally different, these families of compounds are all poorly absorbed by the small intestine, requiring the help of our gut microbes to unlock their health-promoting potential.

Polyphenols benefit the gut in many ways. Some of them serve as prebiotics—that is, as food for our microbes. Some suppress unhealthy microbes in the gut. Most of them are broken down into metabolites that either benefit various cells in the gut connectome or are absorbed into the bloodstream and benefit multiple organs, including the brain. In view of their intimidating numbers and variety, I'll focus on just a few of the essentials.

People who eat berries, red grapes, red apples, plums, and red cabbage and drink moderate amounts of red wine on a regular basis are consuming a lot of *anthocyanins*, a group of molecules within the flavonoid family. They not only contribute to the excellent flavor of these foods, but they're also responsible for their vibrant reds, blues, and purples. Their health benefits have been popularly—but falsely—attributed to their antioxidant effects (more on this in a moment), but in fact, despite their low levels of systemic circulation, anthocyanins and other polyphenols exert their health effects through the gut microbiome.[19] When patients with ileostomies—an external opening of the end of the small intestine created after surgical removal of the colon—ate raspberries, blueberries, lingonberries, and grapes, a large portion of the ingested anthocyanins remained in the ileal fluid, meaning that they'd passed through the small intestine without being absorbed. We now know that the majority of anthocyanins pass intact into the large intestine, where they are broken down into smaller molecules by certain gut microbes before absorption.

The large *flavonoid* family—composed of several thousand distinct molecules—also includes the *catechins*, phenolic compounds present in fruits and berries but more abundantly in cocoa, green (and to a lesser extent, black) tea, and onions. *Isoflavones*, another class of flavonoids, have high concentrations only in legumes, such as soybeans, while oranges, lemons, and other citrus fruits contain a different class known as *flavanones*. These molecules are so varied that a survey of Spanish citrus juices like sweet orange, tangerine, lemon, and grapefruit identified fifty-eight types of flavonoids and related phenolic compounds.

Recent studies of anthocyanin and flavanones shed new light on the complex transformations that occur as they pass through the GI tract. Only trace amounts are absorbed in the small intestine; the majority of ingested anthocyanins reach the large intestine, where they're processed by the microbiota. This yields an array of small molecules that can act on targets in the gut and are also absorbed into circulation, reaching organs throughout the body.[20] Recent research has made it clear that metabolic products of anthocyanins and flavanones are much more abundant in circulation than previously thought.[21] These recent findings not only show the importance of colon-derived phenolic products, but they also refute the false claim made by many supplement producers that polyphenols are primarily antioxidants easily absorbed in the small intestine to reach their targets in the body.

Consequently, I find that when I give talks and lectures, an audience member invariably asks, "When you say polyphenols, are you talking about antioxidants?" There's no question that nowadays the term *antioxidant*—incorrectly meant to describe these molecules underlying the beneficial effects of certain plant-based foods—is used far too loosely among dietitians and diet gurus, causing profound confusion among the lay public. The popular idea behind taking dietary antioxidants is that they'll somehow protect lipids, proteins, and DNA from oxidative damage.

"But, actually, the body is more than capable of keeping in balance its redox status [the balance between oxidants and antioxidants], and it is now well believed that ingested antioxidants represent only a tiny bit of the body's crucial redox regulation system," explained my friend Daniele Del Rio, associate professor and head of the School of Advanced Studies on Food and Nutrition at the University of Parma, Italy, and the scientific director of the Need for Nutrition Education/Innovation Programme of the Global Centre for Nutrition and Health in Cambridge, England. "So don't waste your money loading up on antioxidant pills! And while polyphenols are 'chemically' antioxidants, because of their phenolic structure, their contribution to our health is now thought to be linked to very different mechanisms, completely unrelated to their antioxidant properties."

Despite these emerging details about the fate of polyphenols in our gut, the old antioxidant concept remains deeply ingrained. In fact, as Del Rio told me, "In contrast to the polyphenols, there are other molecules which have *true* antioxidant activity when ingested, including the vitamins tocopherols (vitamin E), carotenoids (which give vegetables and fruits their yellow, orange and red color), and ascorbic acid (vitamin C), which are absorbed way more efficiently in the upper intestine."

However, it's becoming increasingly clear that the nutritional value and health benefits of fruits, vegetables, and their derived products—including MACs and polyphenolic compounds such as flavonoids, phenolic acids, and tannins—are critical, with an estimated 7.8 million premature deaths worldwide in 2013 attributable to fruit and vegetable intake below 800 grams (a little under two pounds) per day.[22] To visualize the recommended daily intake, imagine a platter with helpings of spinach (1.8 ounces), broccoli and cauliflower (3 ounces each), mushrooms (3.5 ounces), sweet potato (8 ounces), blueberries and strawberries (3 ounces each), and half an orange (3 ounces).

It's interesting that the intake of flavonoids varies greatly depending on country or geographical region due to different dietary patterns and differences of assessment methods. The main sources of dietary polyphenols in the US and Europe are coffee, tea, and fruits. The average intake of flavonoids worldwide ranges between 250 to more than 1,500 milligrams per day, including contributions by green and black tea. Perhaps less surprising, the mean intake of total flavonoids in the United States is one of the lowest in the Western world, varying from 250 to 400 milligrams per day. Tea is our main source—as if we needed more proof of our unvaried diet and low consumption of produce.

The highest intake is in Iran (1,650 milligrams per day), followed by the United Kingdom (more than 1,000), with Brazil and Mexico coming in last (less than 150). Populations with a high intake of total flavonoids are those with a high consumption of tea, especially black tea. In Europe, an increasing south-to-north gradient is usually observed; despite the high intake of fruits, vegetables, olive oil, and red wine in Mediterranean countries, the intake of total flavonoids in these countries is lower than in non-Mediterranean countries, as a result of the much higher consumption of tea in those areas. In many ways, it's not surprising that the US comes in at the very bottom of the list, a record mirrored by our minimal fiber consumption. As a culture, we generally tend to take a supplement, pill, or some other product of the medical-pharmaceutical industrial complex rather than take advantage of the natural healing power of our gut microbes.

Nonetheless, there's a different path to take, as there are several familiar foods that contain high amounts of polyphenols in addition to the ones already mentioned—green tea, red wine, and certain spices, such as cloves, cinnamon, turmeric, black pepper, and oregano.

## Green Tea

People have been drinking tea for thousands of years; the earliest direct evidence (tea leaves in the tomb of Emperor Jing) dates to second-century-BC China. Drinking tea is not only an enjoyable social activity with both relaxing and invigorating effects, but it's also associated with many health benefits, including alleviating depression.[23] Both black and green tea come from the leaves of the *Camellia sinensis* plant, but black tea production involves extensive oxidization of the leaves, whereas green tea remains largely unoxidized. While both types of tea are a rich source of flavonoids, the type and amount of this group of polyphenols varies. Green tea contains much more epigallocatechin-3-gallate (EGCG), whereas black tea is a rich source of theaflavins and thearubigins.

Many cell and animal studies have concluded that, in addition to its anti-inflammatory and antioxidative effects, green tea can prevent cardiovascular diseases[24] and may also offer neuroprotective benefits.[25] Even though such claims have never been proven in well-designed human clinical trials, a recent observational study gathered data from thirteen thousand people who took part in the Chinese Longitudinal Healthy Longevity Survey, which provides information about the health and quality of life of the elderly (aged sixty-five and over) in twenty-two provinces in China between 2005 and 2014.[26] This analysis showed that consistent and frequent consumption of green tea was associated with significantly reduced depressive symptoms, especially in men.

There are three well-characterized ingredients that may be responsible for green tea's health-promoting qualities. Tea catechins—primarily the polyphenol mentioned above, epigallocatechin-3-gallate (EGCG)—account for up to 42 percent of the dry weight of green tea; the amino acid L-theanine, 3 percent; and caffeine, 5 percent. These three compounds, alone or in combination, have been shown to make people feel both calmer *and* more alert, with improved memory retention. In addition, EGCG and L-theanine have been shown to calm the brain's stress-response system and lower cortisol levels. They may also play an important role in preventing neuroinflammation and the development of cognitive decline in the elderly.[27]

EGCG is one of the more intriguing compounds. Like most polyphenols, it's too large a molecule to be efficiently absorbed in the small intestine. When the unabsorbed EGCG molecules reach the distal (far) end of the small intestine (also called the *ileum*) and large intestine, they promote the proliferation of beneficial intestinal bacteria, thereby suppressing the relative abundance of potentially harmful microbes and increasing gut microbial diversity.[28] In addition, the bacteria metabolize the polyphenols into smaller molecules that can be absorbed in the distal small intestine and large intestine. These various metabolites are believed to be responsible for tea's reported health benefits on body and brain.

### Red Wine

Despite the well-documented negative effects of regular high alcohol consumption, there's a wealth of evidence from epidemiological studies that moderate consumption of wine has protective effects against several chronic conditions, including cardiovascular and metabolic diseases and neurodegenerative disorders.

It's admittedly in observational studies difficult to tease apart the relaxing effects of alcohol from the healthy outcome of social interactions generally associated with drinking wine. And yet, despite the lack of controlled interventional studies, researchers have mostly attributed wine's health benefits—in particular of red wine—to its polyphenol composition. Although it varies greatly, the amount of polyphenols is estimated to be around 150 to 400 milligrams per liter in white wines and 900 to 1,400 in young red wines. In other words, a pint of red wine might give you most of the daily polyphenol requirement by itself.[29]

Red wine has a unique combination of polyphenols, with the flavonoids as the main group of molecules—such as catechin and epicatechin, tannins, anthocyanins, and flavonols—in addition to such nonflavonoids as stilbenes (including resveratrol) and ellagitannins. As opposed to the alcohol and sugar molecules that are completely and rapidly absorbed in the small intestine, adding often unwanted calories, the polyphenols are targeted at the microbiota in the distal small intestine and colon. A recent review of clinical trials published between 2006 and 2018 examined the effects of red-wine polyphenols and grapes on gut microbiota. Several of these studies reported increased levels of microbial metabolites in feces, urine, plasma, and ileal fluid, confirming the modulation of red wine polyphenols by intestinal bacteria. Furthermore, a substantial body of research suggests that wine polyphenols, similar to green tea, increase populations of beneficial bacteria, while inhibiting growth of pathogenic ones.[30] However, as with dietary fiber, the metabolism, absorption, and circulation of these molecules vary between individuals depending on the microbial community within each.

A study published by Professor Tim Spector and his research group from King's College in London shed more light on this connection. By analyzing data from three different groups of red wine drinkers, they demonstrated that red wine was associated with an increased diversity of the gut microbiota, even in those who drank it only once every two weeks.[31] There was, however, a weaker association of these benefits in the gut microbiomes of white wine drinkers.

What's the alternative for those of us who don't want to drink a glass of wine at all? The study from King's College reported that the relative abundance of the microorganism *Barnesiella* from the phylum Bacteroidetes, which was higher in red wine drinkers, *doubled* in the guts of rats fed black raspberry diets, as shown in a previous study. Furthermore, raspberries have previously been shown to be *four times* richer in polyphenols than red wine.

## Spices

Spices have long been used to add unique flavor and color to cuisines around the world. Indeed, most Indian and other Asian dishes would be unthinkable without their characteristic spices. In addition to their indispensable role in flavoring food, spices such as ginger, turmeric, fennel, mustard, cumin, and cardamom (all of which belong to the same plant family, the Apiaceae) have long been used in traditional Asian healing practices. Turmeric, for example, not only provides the distinctive color and flavor of curry, but it is also considered an effective treatment in traditional Indian medicine for a wide range of seemingly unrelated symptoms and diseases, including asthma, allergies, cough, anorexia, and liver diseases. Similarly, ginger, which was exported from India to the Roman empire more than two thousand years ago, has been used to treat numerous ailments, from colds to nausea, arthritis, migraines, and hypertension. Indians and Chinese are believed to have produced ginger as a tonic root for over five thousand years; in fact, it was considered so medicinally valuable in the Middle Ages that a pound of ginger cost as much as a sheep!

The popularity of these Asian herbs has increased again through natural and complementary medicine, as well as a wealth of research published on their potential usefulness as "antioxidants" in the treatment of cancer, inflammatory conditions, depression, and chronic nausea. It's interesting that many of these conditions are part of the network of diseases that make up today's public health crisis and are linked to chronic activation of the immune system. Unfortunately, the vast majority of this research has been done in test tubes or in cultured cells. Furthermore, most of these compounds, like other polyphenols, don't circulate as easily when taken as a supplement or pill.[32] Conversely, when ingested via foods—given the combinatorial nature of polyphenol composition in plants—these complex molecules exert their beneficial effects on our body and brain with the help of our gut microbes. There are hundreds of related molecules contained in the leaves, roots, seeds, and fruits of the plants from which these spices come. For example, basil leaves contain the polyphenols catechins, quercetin, kaempferol, anthocyanins, and tannins, to name just a few. Other spices with high polyphenol contents include clove, cinnamon, cardamom, coriander, saffron, caraway, black pepper, oregano, and rosemary.

## Olive Oil

The health benefits of extra-virgin olive oil (EVOO) have been reported from preclinical and clinical studies and are applicable to a wide range of metabolic disorders and cardiovascular diseases. EVOO is one of the key health-promoting ingredients of the Mediterranean diet. There are at least two major components that mediate the oil's health benefit—the high concentration of monounsaturated fatty acids (primarily oleic acid) and the high content of polyphenols (primarily oleuropein and hydroxytyrosol).

Polyphenols exert their health benefit with the help of the gut microbiome.[33] Research suggests this may be true for oleic acids, too. Oleic acid is the predominant fatty acid in olive oil—73 percent of its total oil content—while 11 percent is polyunsaturated, such as omega-6 and omega-3 fatty acids. Monounsaturated fatty acids (MUFAs) are quite resistant to high heat, making EVOO a healthy choice for cooking. Traditionally, the high content of MUFAs was considered to be responsible for the protective effects of EVOO, but current evidence suggests benefits are largely related to polyphenols and vitamin antioxidants—vitamins A and E—found in the oil. As many as thirty different polyphenol molecules have been identified in different olives. Furthermore, the phenolic concentration of EVOO ranges from 50 to 800 milligrams per kilogram, and the amount of polyphenols in EVOO depends on the region where the olives were grown, corresponding differences in climate, degree of ripeness when harvested, and the oil-extraction process. In addition, the phenolic fraction of olive oil can vary greatly among different types of olives. As a result, it can be a challenge to figure out which olive oil to buy in order to get the full benefit in terms of both flavor and polyphenols.

I learned more about olive oil a couple of years ago, when I visited my friend Marco Cavalieri, the owner of Le Corti Dei Farfensi in Fermo, on the picturesque Adriatic coast of Italy. In addition to his wines, Marco produces EVOO from eight-hundred-year-old olive trees, using a wide variety of olives, including the Sargano, Carboncella, Ascolana, Coratina, Frantoio, and Moraiolo varieties. (An eight-hundred-year-old tree may sound ancient, but it's practically a sapling in olive oil–making years: olive trees started to grow in the southeastern Mediterranean basin more than six thousand years ago, and they were a major item of trade for the ancient Greeks, Romans, Persians, and Phoenicians throughout the Mediterranean region.) These varieties contain the polyphenols oleuropein, demetiloleuropein, and quercetin, with an average polyphenol concentration of around 800 milligrams per kilogram.

In addition to harvesting the olives from the ancient trees, Marco uses several strategies to ensure the highest possible polyphenol content in his product. The olives are harvested when they have not fully ripened, when their polyphenol production is at its highest. Harvested olives are stored in airtight steel containers to protect them from oxygen and light. Those made into oil are taken to the local facility where they are cold-pressed just hours after they're harvested. The fresh oil has a uniquely pungent flavor and fragrance, with an initial almost burning sensation and taste. In addition to its flavor and health benefits, the polyphenols contribute to its superior oxidative stability compared to other edible oils.

In seeking out the health benefits of the Mediterranean diet, it became clear to me that the high polyphenol content of EVOO makes it a medicine produced by nature and refined by human expertise and traditions. Like any medicine, the precise amount of active ingredients and the quality of processing play major roles in its effectiveness. So rather than being misled by the dark appearance of many expensive olive oils marketed as EVOOs, it is worth investigating where and how they were harvested and processed, as well as their average polyphenol content. This may take a bit of investigating, as most producers don't include information about polyphenol content on their labels. Given the difficulty of tracking down the polyphenol content, the best way for a consumer to determine it is by taste—a pungent flavor is generally a sign of high polyphenol content.

## Omega-3 Fatty Acids

Studies have shown that the two main omega-3 polyunsaturated fatty acids (PUFAs)—eicosapentaenoic acid (EPA) and docosahexaenoic acid (DHA)—offer many health benefits, including prevention of heart disease and cancer and complementary therapy of rheumatoid arthritis, depression, and cognitive decline.[34] While most of these healthy fatty acids are absorbed in the small intestine, there is evidence that they may make it to the large intestine in small amounts, where they increase diversity and change relative abundances.

Foods with the highest concentrations of PUFAs include wild salmon and small fish such as mackerel, herring, sardines, and anchovies; flaxseed, chia seed, and walnuts; and a few other foods, including soybeans, oysters, and cod liver. PUFAs also make up a high percentage of fat in undomesticated animals—such as deer or bison—and their relative content is higher in grass-fed cows compared to conventional farm-raised animals. Despite the fact that PUFAs are found in such high concentrations in these foods, they're also widely used as nutritional supplements—as fish oil and in more concentrated "nutraceuticals," pharmaceutical alternatives that promise physiological benefits. However, like other supplements, controlled clinical studies with omega-3 pills have generally failed to show definitive health benefits.

## When to Eat

There is other useful guidance to be gleaned from the way our ancestors ate. For example, early humans didn't have three regularly scheduled and clock-timed oversize meals, let alone snacks any time of day or night. They also didn't lead sedentary lives or get food delivered to the cave. Takeout and delivery are convenient and safe—perfect for a pandemic—but they've also removed the last reason to exert any physical effort to feed ourselves.

In the Neolithic Age, humans spent the majority of their days hunting, fishing, and foraging. Mealtimes were separated by varying periods without any food but often with a lot of physical activity. I got a sense of this rhythm firsthand while living in a Yanomami village in the Amazon rainforest for several weeks during a documentary film expedition in the early 1970s. These fit villagers' days were jam-packed with activity. They were so busy finding food, they had little time for eating! The women, usually carrying their infants, left the village early in the morning and didn't return until late afternoon, bringing with them the tubers, fruits, and berries they'd collected. The men also went on daylong expeditions, running through the forest in pursuit of game or skillfully maneuvering their dugout canoes through the rapids of the upper Orinoco River. They would all reconvene for dinner in the early evening, go to sleep at sunset, and then wake at sunrise to start the routine again. Essentially, they were practicing a form of time-restricted eating, creating long stretches of up to ten or twelve hours without food, when they were out for the day or sleeping.

Based on an extensive body of preclinical science, it's now understood that regular periods with no food in our GI tracts result in lasting adaptive responses that increase the resistance of our brain-body network to a number of chronic diseases and premature death. Indeed, there *has* been a recent trend toward "intermittent fasting," a catchall term that's come to encompass a variety of diets encouraging reduced caloric intake and/or periods of fasting. In theory, these diets make good use of the practice of fasting, something our ancestors engaged in by necessity.[35]

Fasting—not necessarily keeping the body without food entirely but rather reducing daily food intake—evokes adaptive cellular responses within and between organs, improving signaling pathways whose deterioration is related to metabolic diseases and aging. Fasting maintains the metabolic pathways of glucose utilization by preserving sensitivity to the glucose-lowering hormone insulin and by suppressing systemic immune activation. Ketogenesis occurs. During a fast, carbohydrate-restrictive diets, starvation, or prolonged intense exercise, when no or an insufficient amount of glucose is available, cells are forced to flip a metabolic switch to shift their main fuel source from glucose to ketone bodies, or ketones. These ketones are generated by the liver from fatty acids originating in body fat and distributed to tissues throughout the body, including the brain, as alternative energy sources. Once they reach their targets, they are oxidized in mitochondria for energy. At the same time, cells activate pathways that strengthen the body's defense mechanisms against oxidative and metabolic stress and those that remove or repair damaged molecules. However, ketone bodies are not used just for fuel during fasting; they're also potent signaling molecules with effects on cell and organ functions. It's believed that the highly orchestrated systemic and cellular responses activated during fasting can carry over, even after a person begins eating again, continuing to bolster mental and physical performance and increase disease resistance.

Unfortunately, many of today's "intermittent fasting" diets that make use of ketogenesis require more vigilance and persistence than most of us are able and willing to apply to daily life. These popular regimens include the 5:2 diet, which involves restricting food consumption to 25 percent of your caloric needs two days a week, and the fasting-mimicking diet (FMD), a high-fat, low-calorie diet that helps maintain a physiological fastinglike state without actually having to fast.[36] You may also remember from chapter 5 that there's another form of ketogenic diet that induces ketosis by dramatically decreasing the amount of carbohydrates eaten, often by replacing them with animal fat and red meat. As I mentioned earlier, I do *not* support a regular ketogenic diet, with very few exceptions (such as its use in the treatment of therapy-resistant seizures), for the simple reason that it runs completely contrary to the needs and health of our gut microbiome.

Anecdotally, I've seen many patients and acquaintances give up on these fasting diets, and studies have confirmed that compliance is an obstacle. In the few clinical trials performed with these diets, the most common reason cited by the 25 to 40 percent of subjects who drop out is a lack of motivation to continue fasting. This can also lead to the "yo-yo effect," in which any positive strides toward a normal weight and improved metabolic health are reversed with a relapse, sometimes to a condition worse than at the start of the diet.

Furthermore, while preclinical studies on mice have consistently suggested beneficial effects of intermittent fasting on a wide range of factors—including obesity, diabetes, cardiovascular disease, cancers, neurodegenerative brain diseases, and longevity—*human* clinical studies haven't always shown the same impressive results. In a remarkable illustration of the resilience of the metabolic system, many of these studies suggest that such diets result in weight loss equivalent to standard, low-calorie diets. For example, a 2018 meta-analysis led by Dr. Leanne Harris of the University of Glasgow School of Medicine didn't find significant differences in body weight and fat with intermittent caloric restriction as compared to the traditional low-calorie diets.[37] Other studies comparing several diets based on intermittent caloric restriction have demonstrated that they don't lead to long-term beneficial effects on metabolic and cardiovascular risk profiles in human patients.

## Time-Restricted Eating: Focusing on the Gut Microbiome

Fortunately, time-restricted eating—often wrongly called intermittent fasting—not only returns us to the healthier routines of our ancestors, but is also a more viable option, for it doesn't require reducing overall daily calories. It simply compresses the daily period of time in which we eat. It also takes into account the influence of our circadian rhythms on our microbiomes.

Just a few years ago, the microbiome was thought to be a static community of microbes that, once programmed early in life, would pretty much remain the same until we died. Research has revealed that it is in fact highly dynamic, with daily and seasonal rhythms.[38] In mice, the interactions among the gut microbiota, immune system, and liver show remarkable differences over a twenty-four-hour period. Communication between microbes and our gut occurs much more frequently during a meal, creating profound effects on gene expression in the immune system as well as cells in the gut, liver, and brain. These findings strongly suggest that the timing of our meals can have a profound effect on the gut microbiome and our overall health.

Indeed, two recent mouse studies from the laboratory of Dr. Satchidananda Panda at the Division of Gastroenterology at the Salk Institute for Biological Studies in La Jolla, California, reported remarkable effects of time-restricted eating on metabolism, systemic inflammation, and the gut microbiome.[39] Researchers prevented the mice from access to any food for nine to fifteen hours a day; for the rest of the twenty-four-hour cycle, the animals had unrestricted access to food. In one of these studies, led by Salk staff scientist Amandine Chaix, the researchers found that the benefits of time-restricted feeding were proportionate to the amount of time the mice weren't eating, with fewer advantages observed with less than twelve hours without food. These positive effects also included protection against excessive weight gain when the mice were put on a Western diet high in fat and sugar, even *without* any change in the daily calorie intake. The mice could enjoy their high-fat, high-sugar diet, as long as they ate it in a window of less than twelve hours a day. The investigators also found a reduction in whole-body fat accumulation and associated inflammation, an improvement in glucose tolerance, and a reduction in insulin resistance.

In a second study, led by Dr. Amir Zarrinpar, who now runs the Zarrinpar Lab at the University of California at San Diego, in the restricted group of mice, the normal diurnal fluctuations in the gut microbiome were abolished on a high-fat, high-sugar diet. However, when the mice were allowed to consume the same amount of food on a time-restricted feeding schedule, the rhythmic twenty-four-hour oscillations were restored,[40] diversity increased, and obesity-related microbes decreased. What's perhaps most remarkable about all of these experiments is that, unlike intermittent fasting diets, they didn't involve a reduction in calorie intake but simply a restricted daily feeding time.

These pioneering studies have opened up a new and attractive option to return our bodies to a state of metabolic health. A simple, time-restricted eating strategy can literally let you have your cake and eat it too. You take care of your gut microbial well-being by eating a largely plant-based diet for eight hours of the day and *then* switch your metabolism into ketone-burning mode for the remaining sixteen hours, half during eight hours of sleep at night.

Still, as with any restrictive eating intervention, the major question remains: How realistic is it to think of sustaining this diet in our everyday lives? People around the world have vastly different eating patterns—ranging from eleven p.m. dinners in Argentina and Spain to the late-afternoon dinners of indigenous peoples in the equatorial regions, where sundown is six or seven p.m. Then there are the work-related constraints: a large part of the working population goes to bed early and wakes up at five a.m. for a commute. Many don't get home in time for an early dinner. Schoolchildren need a hearty breakfast before heading out the door and a snack when they get home. Do we really expect people to give up a glass of wine and slice of cheese in front of the TV after a long day? Even with complete control over one's time, it's hard for almost everybody to keep to a rigorous eight-hour-or-less eating window. Reconciling this practice with the rhythms of our lives seems impossible at first glance.

So, in an investigation of my own, and in a show of solidarity, I tried the time-restricted diet myself. Taking advantage of the unique situation created by the COVID-19 stay-at-home orders, my family switched from our three-meal-a-day, traditional Mediterranean diet with snacks (fruits and nuts) to a progressive restriction of the daily "feeding window," beginning with twelve hours and settling ultimately on eight hours each day.

I'll admit: this wasn't an easy shift. We were greatly attached to our old routine, starting every day with a nourishing breakfast while reading the *New York Times*. We were then usually able to reenergize with intermittent snacks, and reliably drank a glass of red wine in the evening before bed. On the weekends, when we typically went out for dinner with friends, one glass of wine often turned into two, and these relaxing, leisurely evenings generally rambled on until eleven p.m.

So we started slowly. Without making any changes to our largely plant-based diet, we began with two weeks of gradually increasing our daily no-eating period from twelve hours to the desired sixteen hours. Once we reached this point, we strictly kept to this daily eight-sixteen rhythm for another month. Think of it this way: you're already (I hope) fasting for half that time while you're asleep.

We arranged it so that we began our sixteen-hour no-eating time no later than eight p.m., skipped our cherished breakfast, and then had our first meal between noon and one. We continued our brisk one-hour morning hikes up to Eagle Rock (on our empty stomachs), a popular point in the nearby Topanga State Park, further increasing the metabolism of our stored body fat as the sole energy source. Surprisingly, and contrary to a common diet myth, our exercise tolerance was not diminished and we didn't experience any symptoms of hypoglycemia. After only two weeks, I noticed that my body weight dropped by two or three pounds a week; this continued for several weeks throughout my self-imposed trial. Although giving up our cherished routine was a big challenge in the beginning, we got used to it, especially when unexpected beneficial habits formed naturally. We automatically dropped our routine of snacking throughout the day (except for the occasional consumption of a fiber bar) and drinking wine after dinner, automatically reducing our daily "hedonic" calorie intake by another five hundred calories per day, while continuing to enjoy our traditional Mediterranean dishes.

After a while, we also allowed ourselves to settle into a more sensible routine. In order to minimize the interference that the strict eight-sixteen schedule can have on people's lives, I recommend, after a month or two, reducing the practice to five days of strict adherence and then two days of habitual food intake. I find it most practical if the unrestricted days fall on the weekend to allow for socializing, late dinners, and full breakfasts.

I'm now into the second month of this schedule and haven't observed any rebound of my body weight, nor has my wife. Rather than continuing its initial weekly drop, my weight stabilized at about twenty pounds lower than when we started, and my wife, who never had a weight problem to start with, settled at ten pounds below her initial weight. By maintaining our daily exercise routine, we prevented any loss of lean muscle mass, as had been reported with some subjects in a recent clinical trial.[41] And, contrary to our initial concerns, we feel less hungry and more energetic on our hikes and throughout the day. Ultimately, our experience has thoroughly convinced me that such a time-restricted eating program is both feasible *and* effective, an observation confirmed by many friends and colleagues. Of course, our experiment wasn't part of a controlled study, and I didn't monitor any biological parameters—so there may have been unnoticed variables, such as weight loss occurring in part due to further unintended restriction of calories—but there's no question in my mind that this approach offers a practical and effective way to improve metabolic health while losing extra weight at the same time.

However, perhaps most important, this approach allows us to optimally feed our gut microbes in a rhythmic fashion—as opposed to the keto diet, which deprives the gut microbiota of essential plant-based ingredients, dietary fiber, and polyphenols. It is important to understand that the time-restricted consumption of a largely plant-based diet is not another short-term weight-loss diet. To reap the most benefit, it should become a lifelong routine, alongside regular aerobic exercise and stress-management practices, such as different forms of meditation. Based on the available science presented in this book, I strongly believe that these changes will not only assure a normal body weight, a diverse microbiome, and a healthy metabolism, but also will protect the brain from the harmful effects of chronic low-grade immune activation.

**Chapter Eight**

# THE KEY TO GUT HEALTH IS IN THE SOIL

In July of 1962, when I was twelve years old, my parents agreed to let me spend summer vacation on my Uncle Johann's farm, in a village some forty minutes north of Munich. This area has long been considered the breadbasket of southern Germany, with large farms stretching for miles, dotted by small towns and villages connected by narrow, winding roads. At the time, these family-owned farms produced a large variety of agricultural products, ranging from wheat and barley to sugar beets and potatoes to milk and meat.

During my visit, we kept a reliable—and reliably arduous—daily routine. We got up at 5:30 a.m. every morning to feed the dairy cows fresh grass and clover, followed by a brief breakfast before going out in the fields again to harvest beets and wheat. At lunchtime, my aunt brought freshly gathered steamed potatoes, which we enjoyed with homemade butter and bread, sitting in the grass next to the wheat fields. (Despite this high-carb diet, nobody in the family was overweight or obese, and nobody had ever heard of gluten sensitivity.) After lunch, we went right back to our strenuous work in the fields. After we'd put in eight hours, we returned home, gathered for a short dinner, went to bed, and quickly fell asleep.

Even though I'll always remember that summer on my uncle's farm as a highlight of my childhood, I never dreamed I would return to agriculture as a part of my professional interest and career. Instead I chose to become a doctor and, about a decade after my summer on the farm, started medical school at the Ludwig Maximilian University in Munich. During my training, I became an expert in diagnosing diseases and treating them with a wide range of medications, but I was never trained to give dietary advice beyond eating adequate protein, carbohydrates, and fat or occasionally restricting some of these macronutrients in certain diseases, such as chronic kidney, liver, or celiac disease. I learned firsthand that successfully fighting illness with the latest, most powerful medications, or at least the ones most heavily promoted by the pharmaceutical industry, has always been the primary goal of the Western health-care system, rather than identifying and preventing its root causes. Furthermore, of all the medications a doctor has in his or her arsenal, antibiotics remain the most effective and successful. They protect us from life-threatening infectious diseases, saving millions of lives.

Nevertheless, we're beginning to see another side to this triumphant story, one that's been unraveling in our guts. Excessive and unnecessary antibiotic use, combined with our modern diet of ultra-processed food, has severely compromised our gut health and played a central role in the long-term global health-care crisis of chronic diseases that underlies the current pandemic.

Meanwhile, an eerily similar story has been unfolding in the environment outside of our bodies. Despite the world's population more than doubling over the last seventy-five years, the rate of hunger and malnutrition has actually *decreased* in many parts of the world because of agricultural innovations that have allowed us to keep pace with our rapidly growing world.

However, we've also been quietly wreaking havoc on our natural surroundings. The health and resilience of our plants, continuously fed with chemical fertilizers, has drastically declined, making them ever more vulnerable to the pests and infections we try to beat back with arrays of toxic sprays. Our inordinate use of antibiotics as growth promoters in animals, raised in overcrowded and inhumane conditions, has led to a decline in the resilience of our entire agricultural ecosystem. We've created a destructive catch-22. As our plants and animals grow more vulnerable to viral infections and diseases, we respond by increasing the use of pesticides and antibiotics to keep everything alive.

The fruits and vegetables in upscale US supermarkets have never looked more appealing, with their bright colors, amazing varieties, and unmarked surfaces, but their decreasing levels of minerals and phytochemicals in their roots bears an uncanny resemblance to the troubles brewing in our dark and oxygen-free guts. Just as the microbiome plays a key role in our health, the microbial ecosystems living in the soil and closely interacting with a plant's root system play a central role in *their* health. Even more stunning, some of the very same molecules essential in interactions between plant-based food and our guts also play a crucial role in the seemingly universal communication system in the rhizosphere—the narrow region of soil covering the plant's root system that root microbiota have chosen as their habitat, also known as the root microbiome.[1]

Even the halolike zone of the rhizosphere is comparable to the outer portion of the gut's mucus layer, the sugarlike coating of the intestinal

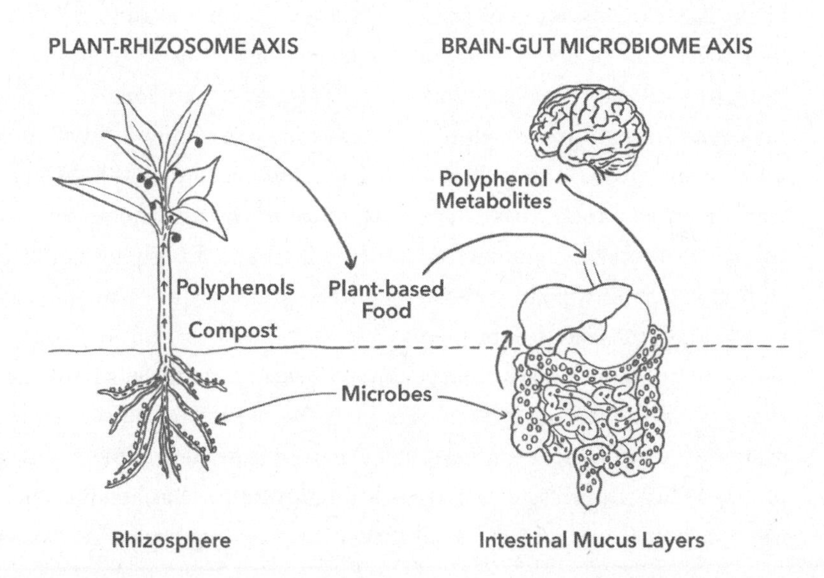

lining where a large number of our gut microbes reside. Likewise, plants produce and secrete a high-complex-carbohydrate fluid that attracts and feeds the rhizosome microbes. And just as the supply of dietary MACs determines the richness and diversity of our gut microbiota, the sugars in this root excretion keep a flourishing ecosystem of soil bacteria alive. Reflecting this similarity in the microbial interactions with our gut and in the rhizosphere, the involved microbes share many of the same genes and metabolic pathways as they turn sugar molecules into energy.[2] The root microbiome contains more microbial life per unit volume than anywhere else in the soil—a dense population akin to the one that exists in our large intestine.

The two systems share a similar plight as well. Under the onslaught of antibiotics, environmental pollutants, an unhealthy diet, and chemical fertilizers and pesticides, the health of our guts *and* the soil microbiome have drastically deteriorated. Our land's devastation has occurred in myriad ways, but one particularly breathtaking illustration is the tallgrass prairie biome, once the most abundant self-sustaining ecosystem in the United States.[3] This area, which stretches from Ohio and Michigan to the eastern Dakotas, Nebraska, and Kansas, was once made up of multiple grass species growing as high as ten feet, in addition to perennials and a variety of wildflowers, covering about 240 million acres of prairie.[4] Between 1800 and 1930, settlers took what they referred to as the Great American Desert (completely missing one of the richest habitats on Earth) and transformed it into farmland, destroying much of this grass prairie and exterminating most of the animals in it, including keystone species like the American bison and the prairie dog. Even more sobering, of course, was the human toll of this expansion, often still shockingly overlooked in history, as European American settlers pushed out or massacred nearly all of the Native Americans.

Agricultural practices then gradually destroyed the very land that the settlers had seized to farm. Plowing the perennial tallgrass root systems led to the near extinction of the prairie grass and many other plant species. Building extensive drainage systems changed the soil's water content and hydrodynamics, resulting in ongoing soil erosion.

This ecosystem once provided abundant food for the native people and some 150 million bison roaming in vast herds, but today it has been reduced to less than 4 percent of its original size. It's been almost completely replaced by monocultures of chemically supported wheat, as well as corn and soybeans to feed enormous numbers of cows.

When I spoke about the health of our soil with Liz Carlisle, who is an organic-farming educator, an assistant professor in the Environmental Studies program at the University of California in Santa Barbara and author of the books *Lentil Underground* and *Grain by Grain*, I asked what led to her passion about the health of our soil. She said, "The beauty that my grandmother experienced in the prairies of western Nebraska and the tragedy of what has happened to lands like that. [My grandmother] was a product of her agrarian childhood; she was so deeply connected to the natural world. And so much of what I loved about who she was came from that connection. . . . She spoke to me quite candidly about the human tragedy that resulted from this failure to care for soil, starting with the genocide of indigenous people. Overall, modern agriculture has dramatically compromised these natural, soil-based ecosystems, with an estimated forty-percent reduction of soil microbiota diversity secondary to tilling and loss of topsoil and chemical fertilizers."

I got my own impression of this hardship a couple of years ago, when I attended a conference in Lincoln, Nebraska, called Microbiomes from Different Habitats—Soil, Water, and Gut. In addition to hearing about experts talking about the decline of microbial diversity in the soil, I had an enlightening conversation with a group of young Native American women from North Dakota. I was curious to learn if they still followed any part of their traditional diet—based on corn, squash, beans, berries, wild rice, and occasional deer or bison meat—or if the ever-expanding influence of Western industrial food had taken over.

"We still eat some of our traditional meals during ceremonial events," one of them told me, "but we no longer have access to many of these foods on our reservation. Our tribe made the decision that it is more profitable to lease our land to big ranches for cattle grazing than to develop our own business based on bison meat or traditional, native foods."

These young women told me firsthand that the prairie ecosystem, once able to completely provide for its indigenous inhabitants, can no longer compete with the massive complex of today's industrial agriculture. The economic equation ignores the hidden costs associated with modern food production. Sadly, two of them told me they'd been diagnosed with metabolic syndrome, yet another illustration of the harsh truth that the economic equation of "cheap food" changes dramatically once the costs of medical care for obesity and metabolic diseases are taken into account.

Meeting these women wasn't the only encounter that brought home for me the ways in which we're failing both our land *and* ourselves. Not too long after the conference, I had another eye-opening experience. I took my wife and son to Bavaria to meet my relatives and see the farm where I'd lived during that transformative summer nearly six decades ago. When we arrived, all of the animals—cows, pigs, chickens, geese, *all of them*—were gone. I was stunned. I tried to reassure myself by looking out at the fields stretching around us for miles, as lush and welcoming as ever. Soon enough, though, my cousin mentioned that cow manure was no longer used to fertilize these fields, as in my youth; instead, they are now treated with a chemical fertilizer. Under economic pressure, the family farmland had turned into a monoculture of corn and winter barley, interrupted only by hop plants—tall, green vines nicknamed Bavaria's "green gold" that have grown here since the eighth century—which have become a valuable commodity in recent times for IPA beers. More than that, all of the hard work that we'd once done in the fields is now entirely accomplished with sophisticated machinery, on a part-time basis by my cousin; the rest of the time he works at a nearby BMW factory. I couldn't help but notice too that several of my relatives, who were once slim and healthy, had gained quite a bit of weight, and many of them were suffering from diabetes and metabolic syndrome.

The relentless dominance of industrial farming practices always results in the same collateral damage of chronic diseases, emphasizing the urgency of this problem on a global scale. For millions of years, evolution has been fighting to optimize the intimate relationship between our microbes and our bodies and between the soil microbiome and plant roots. This insight argues for trying to return to our roots, so to speak, by eating a microbiome-targeted diet, so as to restore our gut *and* plant health.

## Polyphenols: The Health-Care System of Plants

Like the circular conversation within the BGM network, the symbiotic relationship between plant roots and the soil microbiome is a powerful one.

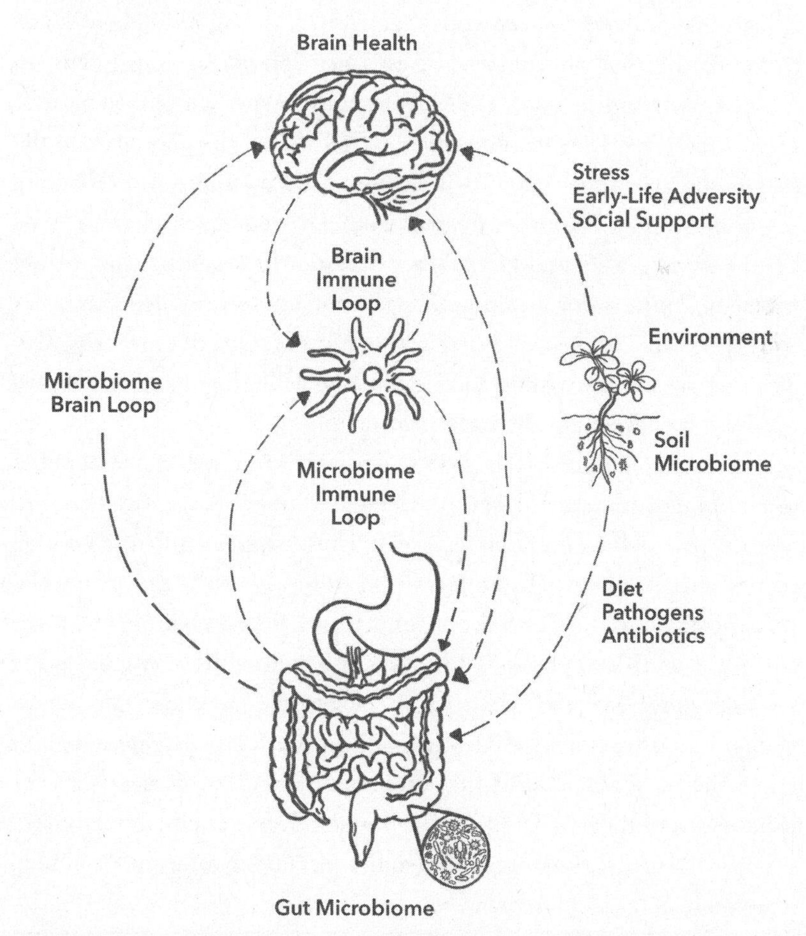

The plants offer free meals of sugars, vitamins, organic acids, and phytochemicals to the soil and in return the soil bestows the plant with a nurturing microbial environment.[5] One example of this mutually beneficial exchange involves polyphenols. One of the largest groups of polyphenols is the flavonoids, abundant in green tea, citrus fruits, berries, legumes, and red wine and among the most important health-promoting elements of a plant-based diet. Flavonoids are abundant in nature; it's been estimated that various plants have more than eight thousand different types.

Flavonoids play a vital role in the plant's health, as they attract nitrogen-fixing species of bacteria that live in or on plant roots and convert nitrogen from the air into a form that becomes a natural fertilizer. Specialized soil microbes, especially of the *Rhizobium* genus living in the root microbiome of legumes (plants that bear fruit in pods, such as lentils, peas, and clover), convert nitrogen into a natural nitrate fertilizer (the amino radical, $NH_2$). When the plant dies at the end of the growing season, it is decomposed by microbes, releasing ammonium ($NH_4$), another form of usable nitrogen, which then feeds neighboring plants and microbes. Growing these plants that attract nitrogen-fixing microbes alongside other crops provides the latter with a natural source of fertilizer.[6] This regenerative process also plays a part in the regulation of atmospheric nitrogen, which makes up about 80 percent of the Earth's atmosphere.

But flavonoids and other phytochemicals offer still more benefits to plants. For instance, when released as root exudates into the soil, other types of flavonoids help render iron, copper, and zinc soluble in the rhizosphere.[7] This allows plants to absorb these minerals, turning them into essential micronutrients for *us* when we consume them.[8] In addition, phytochemicals can be called into action during an emergency, such as when pests, herbivores, nutrient deficiency, drought, or ultraviolet radiation assault a plant.[9] The threatened plant rings chemical alarm bells that turn on an enzyme that catalyzes the production of disease-fighting flavonoids. Simultaneously, the plant sends a distress signal to its rhizosome, recruiting microbes that help it add to its natural pharmacy.

Plants can enlist the root microbiome to fight plant pathogens like the widespread *Pseudomonas syringae*, which can infest plant leaves, stems, buds, and flowers.[10] When the plant detects *P. syringae*, it sends a message down to its roots, stimulating the release of malic acid, which attracts *Bacillus subtilis*, which colonizes the roots and stimulates the plant to produce defensive compounds against the pathogen. *B. subtilis* has also been found in the human intestine, and cultures of it were once popularly used as an immune system stimulatory agent to aid in the treatment of gastrointestinal and urinary tract diseases. Although humans replaced *B. subtilis* with antibiotics some seventy-five years ago, this microbe still continues to play an active role in maintaining the health of plants.

This productive exchange between plants and the soil microbiome extends beyond polyphenols. The essential amino acid tryptophan also plays an important role in the bidirectional interactions between certain gut microbes and the enterochromaffin cells, the serotonin warehouses in our guts.[11] Although tryptophan is broken down by cells lining the gut into several signaling molecules—such as serotonin and kynurenine—most of the undigested dietary tryptophan in the gut lumen is converted by an enzyme that exists only in certain gut and soil microbes into indoles. Indoles have a wide range of functions in the human body and brain. Certain members of this family seem to play a role in ASD, Alzheimer's disease, and depression. In fact, my lab has recently demonstrated that one of its metabolites, indole-3-acetic acid, may play a role in modulating a network in the brain that influences our desire to eat.[12] And just like their cousins in the gut, certain beneficial bacteria in the rhizosome can turn on the same metabolic machinery in the plant and generate the same indole-3-acetic acid metabolite, which turns out to be one of the botanical world's most important growth hormones. In response to this hormone, the plant's roots grow longer, lateral roots develop, and more root hairs sprout. With a more extensive root system, plants can absorb more nutrients and pump more substances into the soil that recruit more microbes helpful in indole-3-acetic acid production.

## The Destructive Duo: The Western Diet and Industrial Agriculture

Though we're only beginning to embrace the notion that our gut and plant health share fundamental underlying principles, all the way down to genes and molecules in our microbiomes, the idea that soil-based microbes play a crucial role in the growth and well-being of plants is an ancient one. As with holistic medicine, this understanding was heedlessly abandoned over the course of history. Since German chemist Justus von Liebig discovered the effectiveness of chemical fertilizer in the early nineteenth century, the prevailing philosophy has been that by fueling plant growth with NPK fertilizer—a mix of the macronutrients nitrogen, phosphorous, and potassium—we can grow an ever-increasing bounty of food.[13] (Incidentally, my cousin now uses NPK fertilizer on his farm in Germany.) Later on, alongside rapid industrialization, Liebig's concept played a major role in the Third Agricultural Revolution, also called the Green Revolution, which improved the output of crops through new agricultural practices in order to address world hunger.[14] This was made possible by high-yielding varieties of cereals (grasses cultivated for the edible component of their grain), as well as the extensive use of an array of new agrochemicals—ranging from massive amounts of NPK fertilizers to chemicals synthesized to kill insects, weeds, fungi, and worms. Together with irrigation, mechanization, and newer ways of cultivation, these practices were promoted as a package to replace traditional farming.

In terms of global food production, it can't be denied that this approach was highly successful. In fact, since 1960, world production of wheat and other grain crops has tripled and is expected to grow further through the middle of the twenty-first century. Rather than basing industrial farming on the biological principles of ecology and microbiome science, however, "Liebig's lasting influence ensured that agricultural science blossomed into a specialized branch of applied chemistry," as David Montgomery, professor of geomorphology at the University of Washington, and Anne Biklé, biologist and environmental planner, put it in their book *The Hidden Half of Nature: The Microbial Roots of Life and Health*.[15] The collateral damage from this modern reductionist approach to a complex system is only now becoming apparent in widespread consequences for the health of our plants, our soils, and our bodies.

Like gut and human health, optimal plant health cannot be had from a simple, cheap chemical mix; there is no such thing that magically makes plants grow, delivering an ever-multiplying abundance of healthy food. No, the sophisticated formula for nurturing our plants occurs naturally in the root microbiome among the vast menagerie of beneficial microbes—bacteria and fungi—that inhabit the immediate vicinity of the plant's roots. This microbiome delivers many of its own metabolites to a plant's roots and also helps the plant absorb minerals, other nutrients, and various otherwise beneficial compounds from the soil. Long before Liebig's discovery built the agrochemical companies, plants took care of their own growth and health in a holistic and highly advanced way, achieving far more than is artificially possible today.

The Western diet delivers easily absorbed high-calorie micronutrient-poor foods while depriving the gut microbes of their preferred nourishment. In exactly the same way, micronutrient-poor chemical fertilizers directly yield mere growth and size of plants while starving the microbes in the rhizosphere. "So we end up with big, fat, high-yielding crops that look good on the outside," Anne Biklé and David Montgomery explained, "but are poor in minerals and phytochemicals on the inside."[16] It's only now dawning on us that our attitudes toward our guts and our soil have severely compromised *both* the complex bidirectional communication between host creatures (all plants and animals, including us) and the estimated trillion species of microbes and their collective wisdom stored in some twenty million genes. And still, despite our phenomenal scientific knowledge, we've uncovered only a thousandth of this genetic intelligence! The NPK diet fueling growth on today's farms doesn't provide our plants with the myriad molecules required to stave off disease, heal from injuries, and fend off pests and pathogens. Just as we've shifted our own diet to one that ignores vital health-promoting functions, so we've also erroneously focused on the growth of our plants while neglecting their inner health. This is where the two failures merge: together they have resulted in an increasingly overweight, obese, and metabolically damaged population requiring an ever-increasing piece of our economy to pay for the pharmaceuticals and medical procedures keeping us alive.[17]

## A "New" Movement: Regenerative Organic Agriculture

Organic agriculture has been one of the earliest ways of cultivating crops. Today it's part of what's known as the organic-food movement, which demands a return to food grown and processed using only natural fertilizers and pesticides. This movement began in the late 1940s, picked up steam in the late 1960s, and is becoming popular enough to be used in marketing by many food companies and restaurants.[18] However, even though organic certification is fairly strict, there *are* loopholes that farmers and agribusiness can take advantage of, so it's possible that consumers are not always getting the benefits they may expect.

For instance, in the United States there are four different categories for organic labeling. "100% Organic" means all of the ingredients are produced organically. "Organic" means that at least 95 percent of the ingredients are organic. "Made with Organic Ingredients" indicates that at least 70 percent of the ingredients are organic. "Less Than 70% Organic Ingredients" requires that three of the organic components must be listed in the ingredients section of the label. It's confusing, to be sure, but "natural" or "all natural" means that the food was *not* produced and processed organically.[19]

While taking us in the right direction by providing fruits and vegetables less covered in pesticides and containing more nutrients, the organic-food movement generally neglects to address the problem of soil degradation. However, a movement has arisen that takes land's health into account—recently enough that I only became aware of it in conversations with a few of its first proponents, including Yvon Chouinard, founder of the outdoor clothing company Patagonia, and environmentalist Liz Carlisle.

"The term *regenerative organic agriculture* is relatively new," explained Carlisle, who collaborates with farmers attempting to restore the ecological integrity of their land, "but it's really a revival of very old practices that have been known to indigenous people around the world for thousands of years."

The return to regenerative agriculture was begun in the late 1980s by the Rodale Institute, a nonprofit that supports organic farming practices. Robert Rodale argued that any properly managed natural system could be productive while increasing its capacity into the future, without relying on expensive and potentially damaging chemical inputs. In short, the farm could be a self-supporting ecosystem.[20]

This concept of holistic land management didn't gain mainstream traction until 2014. The Rodale Institute released a report showing that it's healthier not only for people and land, but also for the planet by helping to counteract climate change. The report concluded, "We could sequester more than 100% of current annual $CO_2$ emissions with a switch to common and inexpensive organic management practices."[21] Essentially, as Carlisle put it, "In the context of contemporary industrial agriculture, regenerative organic approaches basically mean farmers are converting to something that looks more like a healthy indigenous farming system."

Realizing the profound and intricate relationship between soil and gut health allowed me to link my childhood experiences on my uncle's farm and the dramatic changes to it since with the broader shift in how farmers interact with soil and the food they produce. Most important, the realization has opened my eyes to the striking similarities between what synthetic chemical–based agriculture has done to our food and what synthetic drug–based medicine and processed ingredient–based foods have done to our health. It has made it clearer than ever that we must pay attention not only to *what* and *when* we eat but also to *how* our food is grown. Fortunately, there's a strong current of changing public awareness steered by pioneering thought leaders and companies, some of whom I'll profile in the next chapter. They've grasped the enormity of this issue and are striving to help us return our bodies, the land, and the planet to health.

**Chapter Nine**

# THE ONE-HEALTH CONCEPT

Although our health-care system holds fast to the narrative and commercial interests of medical and pharmaceutical companies, there will never be one quick fix for what ails us. Antidepressants alone will not curb the growing numbers of people suffering from depression; the elusive drug sought to slow down the progression of Alzheimer's disease won't remove the underlying risk factors of early cognitive decline; popular regimens such as the keto and low-FODMAP diets provide only short-term benefits for some patients while also degrading the health of the gut microbiome; supplements with megadoses of polyphenols, vitamins, and probiotics won't make up for these severely depleted phytochemicals in our industrially grown plants. Not even the new vaccine for COVID-19 will remove the increasing risk of pandemics transmitted from industrially farmed animals in vulnerable populations. The most pressing public health threats we currently face—chronic diseases, plant and soil health, climate change, and infectious-disease pandemics—are all part of a much larger network out of balance.

Solving these problems in a sustained way, beyond the patchwork of pharmacological, chemical, and complementary remedies

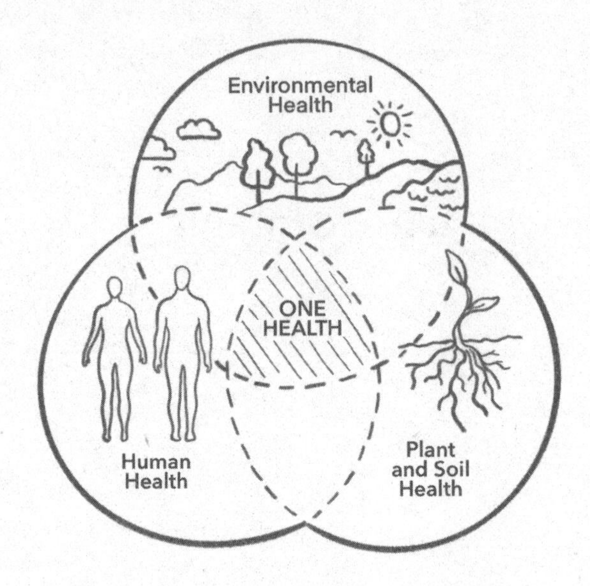


we currently throw at them, will require a new understanding of our planet as an interrelated, systems-based network.

We're only now beginning to comprehend the invisible and largely ignored microbial populations that communicate using myriad molecules in a universal biological language along the various pathways of this global network—from the soil to plants, from plants to humans and other animals, from the gut to the microbiome and brain, and from animals back into the environment. Understanding these intricate relationships is essential to maintaining the health of our organisms, communities, and ecosystems at every level. I believe the idea that there is only one health—the unifying concept of a movement that has historically examined the connection between animal and human health, more recently considering the environment too—should be broadened to encompass a multidisciplinary view of humans, food, microbiomes, animal and plant health, and the environment, with the understanding that these are all imperceptibly connected.

Although one health sounds philosophical and spiritual, in this context it's not theoretical. A paper was published in June 2020 in *Microbiome* by a group of investigators led by Professor Jianming Xu from the College of Environmental and Resource Sciences, Zhejiang

University, in Hangzhou, China, who used a complex network–analysis method to demonstrate a widespread microbial communication system linking animals' intestinal tracts, various parts of plants (including the rhizosphere), soil, and water (fresh and seawater), on a global scale. The researchers established that various seemingly distinct microbial ecosystems are interconnected and communicating with each other.[1]

Investigators analyzed available microbiome data from 23,595 samples and 12,646 exact gene-sequence variants from fourteen environments in the Earth Microbiome Project[2] dataset, an initiative founded at the University of California in San Diego by Rob Knight, who was one of the coauthors of this study. Founded in 2010, the Earth Microbiome Project collects natural samples and analyzes microbial communities from different environments and sites around the globe. The study's "co-occurrence" network analysis revealed eight distinct modules of interconnecting patterns among microbiomes from these various environments. The microbiomes were clustered into two groups, mainly linked by the microbiomes of plant and animal surfaces, such as skin and fur. Although previous findings have observed major distinctions among microbiomes from different environments, further investigation found an overlap of connections among subnetworks, indicating a similarity of microbial patterns among these different environments.

This technique has been widely used to characterize the complex interactions of structure and function within individual microbial communities which determine the role that our gut microbiome plays in our metabolism and cognitive function, as discussed in chapter 4. However, this was the first time it was used to detect interconnected patterns *between* microbiomes of such wide-ranging environments and on such a large scale.

We know from other scientific evidence that over the past seventy-five years there's been a decline in diversity of strains and species at all levels, as well as mismatches between environmental and biological changes, resulting in a widespread decrease in the stability, resilience, and effectiveness of the planetary network.[3] This deteriorating situation is illustrated in a variety of ways: the dramatic decline in biodiversity on macroscopic and microscopic levels, the increase in prevalence of chronic diseases, and a greater vulnerability to pandemics.

Ironically, the SARS-CoV-2 virus, which has caused the COVID-19 pandemic, has exhibited a better understanding of the complexity of our planetary network than the great majority of our doctors, scientists, and politicians. The virus doesn't recognize borders, countries, political beliefs, or the separate organ systems created by medical subspecialties. It knows exactly how to locate and target the most vulnerable people: those suffering from chronic noncommunicable diseases related to the decline of our diet and lifestyle. It also knows about the vulnerabilities that industrial meat production has created for animals and workers living in close proximity under inhumane conditions. It knows about the relentless human encroachment on natural habitats like the rainforests in Africa and South America, increasing its own chances and that of other viruses to jump from compromised animal dwellings to humans. Indeed, this virus almost seems to have its own unique moral compass for what's gone wrong in our modern world—in particular the degradation of our food systems, the economic disparities, and the priorities of profit-driven mega-industries.

While the One-Health concept has long been recognized as legitimate,[4] its potential for delivering meaningful change has now dramatically come to the fore.[5] In 2019, one of the most comprehensive reports based on this approach—exploring the intricate connections among a healthy diet, sustainable food systems, and planetary health—was published in the *Lancet* by the EAT*-*Lancet* Commission, a group of thirty-seven leading scientists from various disciplines and sixteen countries, co-chaired by Johan Rockström, professor of environmental science and codirector of the Stockholm Resilience Center

---

* *E-A-T* stands for *expertise*, *authority*, and *trust*—three factors that Google uses to measure how much trust it should place in a brand or website.

and director of the Potsdam Institute for Climate Impact Research, and Walter C. Willett, MD, DrPH, professor of epidemiology and nutrition at the Harvard T. H. Chan School of Public Health and professor of medicine at Harvard Medical School.[6] As Willett explained when I spoke with him, "This commission was put together . . . to look at the issue of whether and how we would be able to feed a diet that is both healthy *and* sustainable to about ten billion people by 2050." The findings, according to its authors, "provide the first ever scientific target" for reaching this goal.

While the previously mentioned study led by Professor Jianming Xu revealed the crucial role of the microbiome in connecting seemingly unrelated systems on the planet, this report offers an eye-opening reminder of the interrelatedness among food, human health, the environment, and the planet. Even though some aspects of the report have been challenged—a major argument being the impact that its proposed changes in food systems would have on traditional agriculture and dietary habits—it details with great authority and accuracy the inextricable links among modern food systems, poor diet, environmental damage, and human health. Importantly, it ends on the optimistic note that change is possible.

The commission reported that the production and consumption of food in the Anthropocene—our current age, in which human activity has been the primary influence on climate and the environment—represents one of the greatest health and environmental challenges of the twenty-first century. This is not only because the world is dealing with an epidemic of chronic noncommunicable diseases related to obesity and metabolic disorders (currently overshadowed by the more acute COVID-19 pandemic), but also because many environmental systems and processes have been pushed beyond safe boundaries. Disturbances in food supply and consumption have resulted in 2.1 billion adults being overweight or obese and in a doubling of the global prevalence of diabetes in the past thirty years—while at the same time more than 820 million people are undernourished, 151 million children are stunted, 51 million children suffer acute malnutrition, and more than 2 billion people are micronutrient deficient.

Meanwhile, food production is the largest cause of global environmental change. Agriculture occupies about 40 percent of land around the world, and food production is responsible for up to 30 percent of greenhouse-gas emissions. It also absorbs 70 percent of our freshwater use. Conversion of natural ecosystems to croplands and pastures is the biggest factor behind the threatened extinction of numerous species. The runoff from overuse and misuse of nitrogen and phosphorus fertilizers causes dense growth of algae that deplete the water of oxygen and create enormous "dead zones" in lakes and coastal zones. About 60 percent of world fish stocks are exploited and more than 30 percent are overfished. The rapidly expanding aquaculture sector producing farmed seafood can negatively affect coastal, freshwater, and terrestrial ecosystems.

To eat both healthily and sustainably, the EAT-*Lancet* Commission recommends a "win-win" diet, meaning there must be a safe "operating space" for food systems, defined by how much we need of specific food groups daily in order to both maintain human health and the environment—for example, one hundred to three hundred grams (three and a half to seven ounces) of fruit per day.

Willett explained: "We have a lot of evidence for what a healthy diet is. If we look just at health, it points us in the direction of being largely plant-based—not necessarily all vegetarian or vegan but predominantly plant-based. Remarkably, the convergence of scientific advances in a wide range of disciplines, ranging from epidemiology, the microbiome field, metabolism, neuroscience, all the way to plant and soil science, all support the benefits of [this] diet as well."

There is a surprising alternative to giving up on meat altogether for those individuals who are unwilling to dramatically switch their traditional dietary habits. The recent introduction of plant-based meat products and their explosive increase in popularity during the last five years, in particular amongst millennials, is demonstrating that such a dramatic change is possible not only in the US but even in countries like Brazil and Argentina in which daily consumption of beef is part of the national identity. For example, the number of vegetarians in Brazil doubled over a six-year period, which has given rise to a booming plant-based industry that is seeking to replace meatpacking plants and at the same time reduce the devastating environmental impact of the deforestation of the Amazonian rainforest to make room for cattle farms and soybean plantations. Unlike other vegetarian meat substitutes like tofu, the new plant-based burgers are winning over even the most dedicated meat lovers. According to Darren Seifer, an analyst at NPD Market Research, 90 percent of the customers purchasing them are meat-eaters who believe the products are healthier and better for the environment.

It seems obvious that moving away from the high environmental impact of beef production to plant-based ingredients should be good for the planet. For example, the Beyond Burger has about eighteen ingredients, including purified pea protein, coconut and canola oils, rice protein, potato starch, and beet juice extract for coloring. However, it remains to be determined if the beneficial effect for the environment of these ultraprocessed foods is paralleled by individual health benefits. Compared to a beef patty, the Impossible and Beyond burgers have similar amounts of protein, total fat, and calories, with a lower proportion of saturated fat and no cholesterol. While these plant-based products also contain fiber, real meat does not. Studies comparing the metabolic effects of eating beef burgers versus plant burgers are currently under way. According to Dr. Frank Hu, chairman of the nutrition department at the Harvard T.H. Chan School of Public Health, the meat substitutes should be considered "transitional foods" for people who are trying to adopt healthier diets. He cautioned, however, that replacing a hamburger with a plant burger is not an improvement in diet quality if you chase it with French fries and a sugar-laden soda. In August, Dr. Hu, along with a group of health and climate experts, published a report in *JAMA* that explored whether plant-based meats can be a part of a "healthy low-carbon diet."[7] Hu emphasized that replacing red meat with nuts, legumes, and other plant foods has been shown to lower mortality and the risk of chronic disease, but that it is not possible to extrapolate that processed burgers made with purified soy or pea protein will have the same health benefits.[8]

When I asked how large a population our Earth could support if everyone started eating an animal-based keto diet, Willett had a ready answer: "Two hundred million people, which means that about 7.2 billion people would need to find another planet."

The report concludes that the global transformation to such a healthy diet by 2050 will require substantial dietary shifts, including a greater than 50 percent reduction in our consumption of unhealthy foods and a more than 100 percent increase in the consumption of healthy ones. Furthermore, sustainable food production for about ten billion people should use no additional land, safeguard existing biodiversity, reduce water consumption and manage water responsibly, substantially reduce nitrogen and phosphorus pollution, produce zero carbon dioxide emissions, and cause no further increase in methane and nitrous oxide emissions.

These ambitious recommendations come with a stern warning: if current dietary trends aren't reversed or mitigated before 2050, the impact on human health is going to be, as Willett put it, "very direct and severe." The global burden of noncommunicable diseases is predicted to worsen, and the effects of food production on greenhouse-gas emissions, nitrogen and phosphorus pollution, biodiversity loss, and water and land use will further threaten the stability of the Earth. Willett warns us, "It will take efforts of almost everyone making changes in their personal life, as well as policy changes at the national and global levels. It's going to be a huge challenge. It *is* possible to have a healthy and sustainable diet for the world. However, there is not much room for error. We have to move quickly and decisively both in terms of how we produce and consume food, but also in the rest of our lives as well."

This plan is also advantageous from an economic standpoint. The transformation to a healthy diet could effectively deal with our current health crisis, drastically reducing health-care costs and avoiding an estimated 10.8 to 11.6 million deaths per year, which is a reduction of 19.0 to 23.6 percent—more than any pharmaceutical innovation will ever achieve.

As ever, essential questions loom large. Are humans willing to make such dramatic changes to their dietary habits? Or will change occur only as the catastrophes pile up? Will we simply continue to increase health-care and research budgets to deal with noncommunicable diseases and recurrent pandemics on an ad hoc basis? Without immediate answers to such questions, and responding to the realization that there's something drastically wrong with our food production and consumption, a growing number of high-profile companies, farmers, and chefs—revolutionaries, really—have begun to take matters into their own hands. Together they're demonstrating that it *is* possible to grow and produce delicious, nutrient-rich food free of chemicals without harming the environment. Many of them have also created highly successful and profitable businesses based on these principles. From Patagonia owner Yvon Chouinard to Emmanuel Faber, the French CEO of Danone, to chefs and food activists Dan Barber and Norbert Niederkofler, these leaders are illustrating the ways in which we can take back our health on a personal and public scale. Despite their completely different backgrounds, careers, and motivations, they've all converged on a shared basic philosophy aligned with the One-Health principles. They've begun to take practical steps toward radically transforming our food systems.

## The Revolutionaries

Whenever I present Yvon Chouinard, founder and owner of the legendary company Patagonia, as one of the leaders of the One-Health movement, I'm often asked, "Isn't Patagonia the expensive outdoor clothing company?"

Chouinard has a quick and simple answer: "As a lover of the outdoors, I see a way to save our home planet and its creatures—including us—from the destructive habits we've invented. Patagonia Provisions [the division of his business that sells responsibly resourced food] is more than just another business venture. It's a matter of human survival."[9]

I first met Chouinard in 2018 at a speaking engagement in Jackson, Wyoming, where Chouinard lives during summer and fall. We sat in his cozy, modest house, with a breathtaking view of the snow-covered

Teton mountain range. Chouinard spoke plainly and eloquently about his beliefs. "There's nothing wrong with this planet; it's perfect, but we're destroying it, of course. Still, all of the answers are in nature. I think I've always believed that."

Chouinard is passionate about transforming the current model of industrial agriculture to a regenerative organic one as a way of tapping the solutions already held in nature. "Regenerative organic farming practices yield large crops while building healthier soil, which can draw down and store more greenhouse gases," he wrote in a recent essay. "Free-roaming buffalo restore prairie grasslands, one of Earth's great carbon storage systems.[10] Rope-cultivated mussels produce delicious protein while cleaning the water in which they're grown. Place-based and selective-harvest fishing techniques also allow us to target truly sustainable fish populations without harming less abundant species." As these examples illustrate, the more we roll up our sleeves and dig into the world of food, the more we discover that the best ways are often the old ways. Organic agriculture primarily removes all of the harmful chemicals, "but with regenerative organic, you are able to grow more nutritious, better tasting food; you are growing topsoil; and you are capturing carbon from the atmosphere."

Chouinard was fluent—and adamant—about his mission to pioneer a new way of addressing our health *and* environment by changing the way we grow our food. "*That's* the revolution I want to be a part of!"

It's unusual to hear the word *revolution* from a giant of the business world. But then Chouinard is an unusual leader and, in his own words, a "reluctant businessman." His most popular book is *Let My People Go Surfing: The Education of a Reluctant Businessman*.[11]

When I asked Chouinard if he'd ever been invited to speak at the Harvard Business School about his revolutionary and phenomenally successful business model, he replied, "I did give a talk at Harvard to a bunch of MBA students, and one of them came up to me afterwards and said, 'Look, I really enjoyed your talk and I believe everything that you said, but it's the exact opposite of what they're teaching us at Harvard.'" Indeed, Chouinard sees himself more as a socialist than a capitalist and has always pursued what seems morally right and important as opposed to what will make the most profit for his company.

The eighty-year-old Chouinard has ensured that his mission will be widely carried out by creating a program at Patagonia that since its founding in 1985 has offered more than $100 million in grants to grassroots organizations and innovative startups forging new methods of food production in a responsible, regenerative, organic way. Among many other projects, Chouinard has also supported the development of the long-root perennial wheatgrass *Thinopyrum intermedium* into selectively bred grain-bearing strains trademarked as Kernza by turning the grain into several specialty beers, such as Patagonia Long Root Ale. He promotes the consumption of sustainably and humanely harvested salmon and bison meat and leads many educational and marketing campaigns.

With such investments, he's also become an important player in a whole ecosystem of like-minded individuals and companies around the world successfully growing and producing healthy food in a sustainable way. For instance, Emmanuel Faber—the visionary CEO of the international food company Danone, known for its dairy products and bottled water—took on the daunting task of transforming a $30 billion multinational food company with more than a hundred thousand employees into what's called a public-benefit corporation, or a B Corp.* (Patagonia was the first California company to sign up for this status in January 2012.[12]) B Corp status is a sustainable business certification launched in the US in 2006 that requires for-profit companies to have goals that include sustainability, transparency, social responsibility, workers' welfare, animal welfare, and legal accountability.

* While this book was being written, Emmanuel Faber was removed as CEO and chairman of Danone, positions he had held since 1997. His charismatic leadership and his accomplishments in making Danone the biggest public benefit corporation in the world did not prevent shareholder activitists from ending his impactful leadership role there. It was a victory for the activist funds which had laid siege to Danone, pushing for a radical overhaul to fix what they called its "chronic underperformance."

Among other requirements, companies commit to filing a biannual questionnaire measuring social and environmental impact, meeting comprehensive social and environmental performance standards, and making their B Impact Report public.[13]

"We need to rebuild trust as companies, as an industry," Faber stated in an interview. "I think the B Corp certification is a fantastic way of saying to everyone that this is the ethos of the people who are behind the brand. We do not consider the purpose of this company to be returning money to shareholders. There is a broader purpose."

Faber speaks seriously about a complete conversion in the way Danone produces food, moving away from chemicals to crop and animal cultivation based on a deep knowledge of soil, ecology, and climate. In September 2019, he announced the formation of a nineteen-company consortium that pledged to protect the planet's biodiversity through sustainable agriculture and a reduction in deforestation.[14]

"When you look at the food system over the last fifty years, we have been running for scale effects," he told me. "We've been looking for simplification and globalization and focusing on a few solutions. . . . [We] lowered the cost of calories overall, that's for sure. But now the unintended consequence is that we only rely on nine plants, nine varieties, for seventy percent of our food. And this is a huge risk on the ecosystem, because it's an overdependence on only a few species, and we know that these species are not going to be able to adapt."

Following the EAT-*Lancet* Commission's recommendations, Faber agrees that we should eat less meat and dramatically rebalance our diet based on the planet's future health as well as our own. In some cases, this means eating less dairy, Danone's core business, as well as cutting down on or eliminating sugar, which the company has traditionally used in large quantities. To address these issues, Danone has switched to plant-based yogurt and other fermented food, as well as dramatically cutting down on the company's use of sugar. Faber also rejects the use of genetic modification in food production, having pledged to become GMO-free in 2016. Faber doesn't plan to stop when the whole of Danone is a certified B Corp. His vision for the transformation of capitalism is even more radical: "You have a fundamental element of an organization that has a purpose, bigger than itself, which is to serve the interest of a broader ecosystem."

While Chouinard and Faber are driven by a searing ambition to change their respective companies and the world, there are other leaders bringing the movement forward in a more intimate way. Visionary chef and *New York Times* bestselling author Dan Barber unapologetically puts the pleasure of eating at the center of his activism: "I believe that the reason the movement around good food has continued to grow is because . . . it is rooted in hedonism. We can bemoan the American food culture and our lack of appreciation for good food—but one thing that we have more than anyone else is a culture that is greedy for pleasure and willing to go out of their way for it and spend a little bit more money [for it]. And *that* is where I think this movement will continue to have legs."

I agree with Barber in every way, including from a scientific perspective. As I pointed out in chapter 6, our motivation to eat is driven by two balanced biological drives, one being the metabolic needs of our bodies and brains, and the other, as Barber instinctively picked up on, is hedonic. Pleasure is a huge driving aspect of our eating behavior. While this subject is of intense scientific and commercial interest—both by food companies that add flavors to their products with the hope that it will make us crave them all the time *and* by the pharmaceutical industry's race to make a pill that will reduce food intake—the pleasure-seeking aspect of our biological drive can also play a crucial role in urging us toward eating healthier, tasty foods.

With Michael Mazourek, Dan Barber founded and operates Row 7 Seed Company, which sells organic seeds. Barber uses a flavorful carrot to illustrate the complex behind-the-scenes agricultural process required for the kind of healthy and delectable meal that keeps us coming back for more. "[That carrot] has to start with fair seed, probably a seed that was selected with care and consciousness over many, many years or generations. It was probably grown from a local farmer, and it probably got to you pretty soon after it was picked out of the soil. And by the way, that soil . . . was real soil and not a hydroponic area because the taste—you don't get that hydroponically. So you had to have soil that was teeming with life, full of nutrients, biologically diverse and loaded with fertility, because that's where flavor comes from. Also, soil doesn't work in a vacuum. It only works where the entirety of the farm is in play. There is, for example, good pollination and good bird life, things that make the environment robust and strong. And the nutrient density of that carrot, well, that had to be off the roof because the flavonoids and other polyphenols are essential for flavor. So this delicious carrot you just tasted was a nutrient bomb combining both nutrient density *and* deliciousness."

Barber isn't finished with his analogy; he's just getting started. "I consider this a political carrot right there. That's a carrot with all of those attendant issues attached to it to become a prism, [filtering] light onto decisions that determine how the world is used. So actually taste plays a very big role in . . . how we might think about the future of food."

Some six thousand miles east of Barber's New York State restaurants, high up in the Dolomite Mountains of Northern Italy, lives another like-minded chef. I met Norbert Niederkofler, of the renowned Restaurant St. Hubertus at the Rosa Alpina Hotel in San Cassiano, in the Italian Alps, at the beginning of 2020. Along with Chouinard, I'd been invited to speak at CARE's Ethical Chef Days Conference, cofounded by Niederkofler and his business partner Paolo Ferretti. It's an annual event, gathering chefs, other food professionals, and executives of wine and food companies from all over the world who share the same objective: to advocate for an ethical and sustainable approach to cuisine, recognizing the responsible use of raw materials, the hard work of farmers and breeders, and the recycling of waste.

I confided in Niederkofler that we'd met once before some fifty years ago, when he was a culinary attraction in a first-class restaurant on a lake near my hometown in Bavaria. My dad took our family there for a special dinner. What a coincidence that we met again half a century later, after both of us had traveled the world, only to converge on the same philosophy—Niederkofler from a gastronomical angle and I from the perspective of health.

St. Hubertus was founded in 1996, beginning as a small restaurant inside the hotel pizzeria, with Niederkofler as head chef. Four years later, it was awarded a Michelin star, the first one ever received in the Dolomites. "We flew in products from all over the world—from Australia to Alaska to Norway," Niederkofler recalled. "Because the restaurant was famous for seafood, we had 150 kilos of sea-fresh fish flown in every week." When the restaurant received a second Michelin star in 2007, he said, "I began to realize that it didn't make sense any more to bring people from all over the world to the mountains to serve them the same food as they can have in New York and Los Angeles, in Tokyo and in Australia."

Niederkofler began "cooking the mountain," as he says, sourcing everything nearby. "A lot of journalists and even the officials from the Michelin organization warned me that if I am going on with this way, I would lose my second star. But I knew that I couldn't continue the old way, where we were just not taking care of the world and the environment. All the wisdom I had learned when I was living with the Hopi Indians told me it is absolutely the wrong way."

Niederkofler's new cuisine varies with the Dolomites' seasons and is prepared according to centuries-old cultural traditions of the local farmers. His approach is deeply rooted in respect for the environment and the importance of close social interactions. In fact, Niederkofler has built a local supply chain of around fifty farmers. He visits them on their farms. "You keep the culture, and you understand why they have done [it this way] in the old days . . . so you can do really good and honest storytelling" on menus and in conversations with his guests. While cooking, Niederkofler forgoes olives or citrus fruits because they do not grow at the high altitude of the Dolomites. His menus are printed on paper made from apple pulp. He never uses vegetables out of greenhouses, and he avoids leaving any waste.[15] "Nature gives you exactly what your body and your mind need in the colors in every period of the year," Niederkofler says. "Nature decides [our menu] for us because it gives us the products when they are ready."

In 2017, St. Hubertus was awarded a third Michelin star.

It's obvious that changing the trajectory of our food system is closely connected with slowing or reversing climate change and transforming the way we take care of our health. Accomplishing this change is a colossal task, which will take the combined efforts of dedicated individuals in all industries and demographics—consumers, patients, scientists, business owners, food-industry leaders, and politicians included. The stakes have never been higher or the job more urgent. We must simultaneously combat the public health crisis, all-but-certain future pandemics, catastrophic climate change, and their devastating effects on billions of people all over the world. None of these efforts will succeed without a fundamental change in the general human mind-set at the individual level.

When we select food at the grocery store, it's essential that we consider how and where it was produced, what impact its production had on the farm workers, the environment, and the climate, as well as how beneficial it is for our gut microbiome and ultimately for our bodies and brains. We need to adopt a more holistic understanding of the complex networks of life that exist at multiple scales—from the microbial networks in our gut and soil to the brain-gut-microbiome network and ultimately to our entire planetary network. If global change to a healthy food system seems impossible, remember that these combined efforts will offer us the possibility of living into our eighties and nineties *without* relying on the medical-pharmaceutical industrial complex and will also maintain the larger ecosystem of our planet to which we're inevitably bound.

**Chapter Ten**

# A NEW PARADIGM FOR HEALTHY EATING

After detailing the wide-ranging health benefits of a largely plant-based diet throughout this book, I've planned this last chapter to make this knowledge practical and put it into action in the kitchen.

When we focus primarily on the nutrients essential for gut and microbiome health, we automatically provide our body with adequate high-quality macro- and micronutrients. When we feed our microbes a diverse menu of plant-based fiber and polyphenols, we don't have to worry about getting enough protein, vitamins, and minerals. At the same time, when we focus on eating foods that are good for the gut, we tend to eat fewer unhealthy foods and fewer calories.

Here you'll find an assortment of recipes for every meal of the day. Some are more elaborate; others are quick-assembly meals, like bowls and smoothies. For each of them, I've included a chart that ranks the meal based on different nutritional criteria than we're accustomed to seeing in cookbooks or on food labels. I've assigned each dish a healthy-food index (HFI) score assessed per serving based on the content of microbiome-accessible carbohydrates (MACs or fiber), polyphenols, omega-3 fatty acids (all in grams), and calories. The HFI is calculated as follows: [MACs + polyphenols + omega-3 fatty acids] x 100 ÷ [calories + absorbable carbs (total carbs - MACs)]. A list with the macronutrients total fat, protein, and total carbs together with details about the individual polyphenol content of each dish can be found on my website emeranmayer.com. Consider a standard lunch of a cheeseburger, French fries, and a Coke. The HFI is 0.62, and it is about the same for a prime rib steak with a baked potato (0.73). Compare this with the HFI of the first recipes in this chapter: Wild Rice with Mushrooms (2.74) and Moroccan Lentil and Chickpea Stew (3.04). And don't forget that the health benefit of a largely plant-based diet is determined not only by the total *amount* of plant-based food, but also by its *variety*. The greater the variety, the greater the diversity and richness of gut microbes that are being nurtured. For additional, detailed information about the nutrient composition of these recipes, please visit my website: emeranmayer.com.

Once you understand the way we can quantify the health-promoting effects of our diet, you can customize these recipes to suit your tastes and perhaps even be inspired to create new ones. For example, in addition to whole-wheat pasta, there are many types of nonwheat pastas that are high in fiber, low on the glycemic index, and sustainably produced. Likewise, small fish, such as mackerel, sardines, or anchovies, can be substituted for fish higher on the ocean's food chain, or you may choose sustainably harvested wild salmon.

Some of the recipes in this book were contributed by Orsha Magyar, CEO and founder of NeuroTrition Inc., by Annie Gupta, PhD, and by Chef AJ, and several were selected and modified from publicly available recipes, including Patagonia Provisions. These sources are indicated at the end of each recipe and on the Resources page.

I've also included sample meal plans to help guide you, but I realize that switching to a time-restricted eating schedule can create some challenges and may require changes in long-ingrained habits. Based on personal experience and discussions with many individuals who have successfully switched to a schedule that restricts food intake to an eight-hour window, it seems that one of the easiest ways to do this is to refrain from eating after eight or nine p.m., and delay your first meal the next day until noon or one. If you prefer to have dinner at six and then eat brunch at ten a.m. the next day, that's fine too. It's important to make time-restricted eating work for your life and the demands of your schedule. Generally speaking, though, you will eat two meals a day, with healthy snacks in the form of fruits, vegetables, or fiber-and-nutrient bars in between meals if needed. For many people, the greatest challenge of this schedule is giving up a round of snacks and drinks after dinner. In order to stay compliant over the long term, and to give yourself a break to be social and enjoy special meals with friends and family, I recommend returning to a regular eating schedule on the weekends, after you've completed your first month of time-restricted eating.

Always remember, switching to this style of eating is not a temporary fix to combat overweight and obesity, only to be replaced by the next trend. Our Western diet not only makes us fat, but it is also one of the root causes of our entire public health crisis. The real problem with it is not the extra pounds, but the long-term consequences of metabolic dysregulation, including cardiovascular disease, cancer, cognitive decline, and greater vulnerability for infectious diseases. To return to and strive for optimal health requires a fundamental lifestyle change based on the realization that the health of our bodies, our gut, the plants, and the environment are all nodes in a gigantic network that depends on the integrity of the microbial world inside and outside us. To repair this global network, we must commit to making healthier choices for a lifetime.

# Recipes

## Mains*

### Wild Rice with Mushrooms

| Healthy Food Index | Calories | MACs | Polyphenols | Omega-3 FA/ Total Fats |
|---|---|---|---|---|
| 2.78 | 427 | 6.7g | 2.477g | 0.188 |

(listed amount per serving)

**SERVES 4**

1½ cups wild rice
2 tablespoons extra-virgin olive oil
1 medium leek, split lengthwise and sliced thinly
1 yellow onion, sliced
Salt and pepper to taste
1 teaspoon turmeric
1 teaspoon thyme
2 cups sliced mushrooms (shiitake, morel, or portobello)
¾ cup walnuts, roasted and crushed into smaller pieces
2 teaspoons capers *(optional)*

Cook wild rice according to instructions on the package.

In a skillet, heat olive oil on medium heat and sauté leek and onion until tender, about 7–8 minutes.

Add salt, pepper, turmeric, and thyme.

When herbs are fragrant, add the mushrooms to the pan and cook for 4–5 minutes, until soft.

Add walnuts and continue to cook on low heat for another 2–3 minutes.

Stir in the cooked rice, add capers if desired, and continue to cook for another couple of minutes to allow flavors to meld. Remove from heat and serve.

Contributed by Minou Mayer, MA

* Calories, MACs, polyphenols and omega3 are all per serving.

# Moroccan Lentil and Chickpea Stew with Candied Walnut and Chia Seed Garnish

| Healthy Food Index | Calories | MACs | Polyphenols | Omega-3 FA/ Total Fats |
|---|---|---|---|---|
| 3.04 | 547 | 17 | 0.117 | 0.19 |

(listed amount per serving)

**SERVES 4–6**

- 2 tablespoons extra-virgin olive oil
- 1 yellow onion, finely diced
- 4 garlic cloves, minced
- 2 teaspoons cinnamon, ground
- 1 tablespoon cumin, ground
- 1/2 teaspoon red chili flakes
- 1 teaspoon coriander, ground
- 1/2 teaspoon cloves, ground
- 2 teaspoons dried ginger, ground
- 1 1/2 teaspoons sea salt
- 1/4 teaspoon black pepper
- 5 cups cauliflower, cut into bite-size pieces
- 7 cups spinach
- 3/4 cup sulfite-free dried apricots, diced
- 28-ounce can diced tomatoes
- 4 cups low-sodium vegetable broth, or water
- 1 cup green lentils
- 18-ounce can chickpeas, drained and rinsed
- Optional garnish: chopped cilantro or other polyphenol-rich herb, like parsley

In a large sauce pot heat olive oil over medium heat.

Add onion and garlic and cook for 3–4 minutes, until onion is translucent.

Add cinnamon, cumin, red chili flakes, coriander, cloves, ginger, salt, and pepper, and let cook for 3 minutes. Then add cauliflower, spinach, apricots, diced tomatoes, vegetable broth, and lentils.

Bring to a simmer and cook until the lentils are tender but still firm, about 45 minutes.

Add the chickpeas and cook for 5 more minutes. Ladle the stew into bowls and top with Candied Walnut and Chia Seed Garnish (see next page) and optional cilantro.

Contributed by NeuroTrition Inc.

# Candied Walnut and Chia Seed Garnish

| Healthy Food Index | Calories | MACs | Polyphenols | Omega-3 FA/ Total Fats |
|---|---|---|---|---|
| 3.26 | 281 | 3 | 0.047 | 0.24 |

(listed amount per serving)

**SERVES 4–6**

**1 tablespoon extra-virgin olive oil**
**1¼ cup raw walnut pieces**
**½ teaspoon dried ginger, ground**
**½ teaspoon true cinnamon, ground**
**1½ tablespoons pure maple syrup**
**½ tablespoon chia seeds**
**Pinch of sea salt**

Heat olive oil in a medium to large pan over medium heat. Add walnuts, ginger, and cinnamon and stir to coat.

Drizzle the maple syrup on the walnuts, then sprinkle the chia seeds and sea salt on top. Heat on stovetop for 5 minutes, stirring often to prevent burning.

Remove walnuts from the pan and let cool for 5 minutes on a baking sheet or in a glass baking dish. Once cooled, break apart pieces that are stuck together.

Contributed by NeuroTrition Inc.

## Seared Sea Bass with Cauliflower "Rice" Pilaf

| Healthy Food Index | Calories | MACs | Polyphenols | Omega-3 FA/ Total Fats |
|---|---|---|---|---|
| 1.62 | 378 | 5.3 | 0.08 | 0.03 |

(listed amount per serving)

**SERVES 4**

FOR THE CAULIFLOWER "RICE" PILAF:
1 tablespoon olive oil
1/4 cup red onion, diced
2 teaspoons garlic, chopped
1 teaspoon black nigella seeds or cumin seeds
1/3 cup carrot, diced
3 cups fresh cauliflower, chopped
1/3 cup low-sodium vegetable broth
1/2 cup canned artichoke hearts, quartered
1 cup kale, stems removed, chopped into bite-size pieces
1/4 cup raw almonds, chopped
1/4 cup fresh coconut, shaved
2 tablespoons fresh herbs (chives, thyme, parsley . . .), chopped
Sea salt and pepper to taste

FOR THE SAUCE:
1 teaspoon coconut oil
1 teaspoon each fresh ginger and turmeric, finely chopped
1/4 teaspoon chili flakes
1/2 cup sliced plums
1 teaspoon tamari
1/2 cup unsweetened coconut yogurt alternative
Sea salt and pepper, to season

FOR THE FISH:
4 five-ounce pieces sea bass
Sea salt and pepper, to season
2 teaspoons extra-virgin olive oil
Optional garnish: hemp hearts, broccoli sprouts

In a medium saucepan heat oil on medium heat. Add the onions, garlic, and black nigella seeds and sauté for 5 minutes, until fragrant.

Add the carrots and reduce heat to low, then cook for 5 minutes to start softening the carrots.

Add the chopped cauliflower and sauté for 5 minutes, then top with vegetable broth and simmer for 5 minutes until the vegetable broth has reduced and the vegetables are tender. Set aside while preparing the fish and sauce.

To make the sauce, warm the coconut oil in a small pot until melted, add the ginger, turmeric, and chili flakes, and heat until fragrant. Add the sliced plums and tamari, and heat through to allow the plums to release their juice, about 3–5 minutes.

Stir in the coconut yogurt and season to taste with salt and fresh pepper. Cover and set aside.

Season the fish fillets with salt and pepper, and heat oil in a frying pan until it is shimmering.

Place the fish gently in the pan and cook on medium heat for 8 minutes, then flip the fish and continue to cook for another 8 minutes (or until fish is firm and flakes when touched).

To finish, rewarm the cauliflower rice, and add the artichokes, kale, almonds, and coconut. Heat until the kale is wilted.

Add the fresh herbs, and season to taste with salt and pepper.

To plate, spoon the "rice" pilaf on the plate and top with the sea bass, then spoon sauce onto the fish. If using hemp hearts and broccoli sprouts, add on top of the sauce right before serving.

Contributed by NeuroTrition Inc.

# Reinvented Shepherd's Pie

| Healthy Food Index | Calories | MACs | Polyphenols | Omega-3 FA/ Total Fats |
|---|---|---|---|---|
| 2.75 | 341 | 10 | 0.088 | 0.006 |

(listed amount per serving)

**SERVES 4**

2 Asian sweet potatoes
1 tablespoon extra-virgin olive oil
1 teaspoon fresh ginger, grated
1 teaspoon garlic, chopped
1/2 cup red onion (1/2 large onion), chopped
1/2 cup carrot (1 medium carrot), diced
1 cup butternut squash, diced
1/2 cup button mushrooms, quartered
1/2 teaspoon Korean chili flakes
1 tablespoon chickpea miso paste
1 1/2 teaspoon Korean chili paste
1 cup low-sodium vegetable broth
1 cup Napa cabbage, chopped
1 cup Japanese eggplant, diced
1/2 cup plant-based kimchi, chopped into bite-size pieces
1/2 cup raw walnut pieces
1 cup cooked lentils
3 baby bok choy, cut lengthwise into quarters (sixths if large)
1 green onion, cut into 1-inch lengths
4 teaspoons tamari, divided
1/2 teaspoon ginger
1/4 teaspoon sesame oil
1 teaspoon sesame seeds
Optional garnish: black sesame seeds, hemp hearts, chopped chives

Preheat the oven to 350°F, score the sweet potatoes, and roast for 1 hour until soft while making the filling for the shepherd's pie.

Heat oil in a pot on medium heat, and sauté ginger, garlic, and onion until fragrant.

Add and sauté the carrots, squash, and mushrooms for 5 minutes.

Add the chili flakes, miso paste, and chili paste. Cook for 5 minutes to release the flavors.

Add the vegetable broth and simmer for 10 minutes, until the vegetables start to soften.

Add the cabbage and eggplant and continue to cook until softened.

Turn the heat to low and add the kimchi, walnuts, cooked lentils, bok choy, and green onion, simmering for 10 minutes until soft.

Taste and season with 2 teaspoons tamari (or to suit your taste).

Divide between four 10-ounce ovenproof casserole dishes, and set aside while preparing the sweet potatoes.

Pull the skin off the sweet potatoes and mash with a fork, then season with the rest of the tamari and the ginger, sesame oil, and sesame seeds.

Divide the sweet potato mixture evenly onto the stew mixture and place back in the oven to bake for 10 minutes and warm everything through. Garnish with black sesame seeds, hemp hearts, and chives.

Contributed by NeuroTrition Inc.

## Pasta Dishes

# Pasta con Sarde

| Healthy Food Index | Calories | MACs | Polyphenols | Omega-3 FA/ Total Fats |
|---|---|---|---|---|
| 3.72 | 458 | 16 | 0.135 | 0.06 |

(listed amount per serving)

**SERVES 4**

8 ounces edamame spaghetti
3 tablespoons extra-virgin olive oil
1 large onion, finely chopped
1 fennel bulb, finely chopped
1 tablespoon chili pepper
1 tablespoon fennel seeds
8 ounces sardines, boneless and skinless
2 cups tomatoes, sliced
6 ounces white wine
1/2 cup water
1 tablespoon thyme
Salt and pepper to taste
2 tablespoons chives, finely chopped

Cook pasta according to the instructions, drain, and set aside.

Heat olive oil over medium heat in a large skillet.

Add onions, fennel, chili, and fennel seeds and cook about 5 minutes, until onions have reduced.

Add half of the sardines and stir well.

Add tomatoes into the mixture.

Add wine and 1/2 cup water.

Sprinkle thyme, salt, and pepper, and stir.

Add the rest of the sardines, and simmer on low heat for 7–8 minutes, until most of the liquid has reduced.

Transfer cooked pasta to the pan with the sauce and stir gently to coat. Sprinkle with chives and serve.

Adapted from *Jamie's Italy* by Jamie Oliver

# Cauliflower Fettuccine Alfredo with Homemade Cashew Cheese

| Healthy Food Index | Calories | MACs | Polyphenols | Omega-3 FA/ Total Fats |
|---|---|---|---|---|
| 2.06 | 478 | 10.7 | 0.03 | 0.004 |

(listed amount per serving)

**SERVES 4**

- 8 ounces whole-grain (gluten-free, if desired) fettuccine or linguine noodles
- 4 cups steamed cauliflower
- 1 garlic clove, whole
- 1 teaspoon onion powder
- ½ teaspoon sea salt
- Pepper to taste
- 1 cup homemade Cashew Cheese (see next page), or store-bought
- ½ cup plant-based milk
- 3 tablespoons extra-virgin olive oil
- 1 cup frozen peas
- Optional garnish: fresh parsley, raw hemp hearts, pepper

Bring a large pot of water to boil and cook pasta as directed on the package, until al dente.

While the pasta is cooking, make your Alfredo sauce by blending the cauliflower, garlic, onion powder, sea salt and pepper, Cashew Cheese, plant-based milk, and 2 tablespoons of the olive oil in a blender. Blend on high speed until creamy and smooth. Adjust seasoning if needed.

Once the sauce is done, heat 1 tablespoon olive oil over medium heat in a medium-size saucepan. Add peas and cook for 3–5 minutes until warm.

Add the cauliflower Alfredo sauce to the peas, then toss the noodles in the sauce and serve immediately.

Garnish with chopped parsley, hemp hearts, and freshly cracked pepper if desired.

## Homemade Cashew Cheese

**MAKES 8 OUNCES**

- 1 cup raw unsalted cashews
- 2 cups boiling water
- 1/4 cup filtered, room-temperature water
- 1 probiotic capsule

Put cashews in a glass bowl and fully cover with boiling water. Soak for 2 hours, then drain and rinse.

Transfer cashews to a blender and blend at high speed until smooth, adding approximately 1 tablespoon of room-temperature filtered water at a time to help achieve a smooth consistency.

Break open the probiotic capsule, sprinkle the contents into the blended cashews, and blend one more time to incorporate throughout.

Transfer to a glass bowl, cover with a towel, and place it in the oven with the light on overnight.

**Note:** This cashew cheese will last for one month stored in an airtight container in the fridge.

Contributed by NeuroTrition Inc.

# Pasta del Frutta di Mare

| Healthy Food Index | Calories | MACs | Polyphenols | Omega-3 FA/ Total Fats |
|---|---|---|---|---|
| 2.63 | 354.5 | 8.7 | 0.017 | 0.062 |

(listed amount per serving)

**SERVES 4**

8 ounces whole-wheat spaghetti
2 tablespoons extra-virgin olive oil
2 cups leeks (light green and white parts only), chopped
2 medium bell peppers, cleaned and cut in small slices or squares
2 cloves garlic, minced
1 1/2 cups cherry tomatoes, halved
2 tablespoons chopped herbs, such as cilantro, thyme, and chives
2 tablespoons capers, rinsed and drained
1 lemon, zested
1 can smoked mussels
3 ounces smoked wild salmon, cut into small pieces (discard juices and skin)
Salt and freshly ground pepper

In a large pan, boil water and cook pasta according to package directions. Drain and put aside.

In a large frying pan over medium heat, heat 2 tablespoons of olive oil.

Add chopped leeks and bell peppers and a pinch of salt and cook 3–4 minutes.

Stir in minced garlic and continue to sauté about 2 minutes on medium heat.

Add tomatoes and herbs and cook until the tomatoes soften, 3–4 minutes.

Stir in capers and lemon zest.

Add cooked pasta to the pan and gently stir to coat.

Add mussels (with juice) and salmon to the pasta.

Continue to cook for another couple of minutes.

Season as desired and enjoy.

Contributed by Minou Mayer, MA

# Spaghetti al Funghi

| Healthy Food Index | Calories | MACs | Polyphenols | Omega-3 FA/ Total Fats |
|---|---|---|---|---|
| 2.18 | 268 | 5.9 | 0.07 | 0.013 |

(listed amount per serving)

**SERVES 4**

1 medium spaghetti squash (seeded)
2 tablespoons extra-virgin olive oil
1 medium yellow onion, sliced
1 small leek, thinly sliced (white and light green parts only)
1 tablespoon ginger
1/2 teaspoon cloves
1 teaspoon lemon pepper seasoning
1 teaspoon chili pepper *(optional)*
1 1/2 cups mushrooms, sliced
1 1/2 cups fresh tomatoes, sliced
2 cups broccoli florets, chopped into small pieces
1/2 cup toasted pumpkin seeds
Salt and pepper to taste

Cut the spaghetti squash in half, remove the seeds, and wrap each half in a paper towel.

Place the two halves in the microwave and cook for about 8 minutes.

When done, leave in the microwave for 5 minutes to cool before removing.

While the spaghetti squash is cooking in the microwave:

In a pan, heat olive oil on medium heat and add the onion and leek.

Sauté until the onion is translucent, about 5–7 minutes.

Add spices (ginger, cloves, lemon pepper, chili pepper) until fragrant.

Add mushrooms and sauté until tender, about 3 minutes.

Stir in tomatoes and broccoli and cook on medium-low heat 5–6 minutes, until broccoli is tender.

Using a fork, add the spaghetti squash to the vegetable mixture and gently toss together.

Add salt and pepper to taste.

Sprinkle toasted pumpkin seeds on top and serve.

Contributed by Minou Mayer, MA

# Walnut Dukkah-Crusted Salmon with Lemony Kale

| Healthy Food Index | Calories | MACs | Polyphenols | Omega-3 FA/ Total Fats |
|---|---|---|---|---|
| 3.5 | 264 | 2.8 | 0.04 | 0.32 |

(listed amount per serving)

**SERVES 4**

**1/4 cup raw walnut pieces**
**1/2 teaspoon each pepper and sea salt**
**2 teaspoons cumin seeds**
**1 teaspoon fennel seeds**
**1/4 teaspoon coriander powder**
**3 tablespoons sesame seeds**
**4 six-ounce wild salmon fillets**

Preheat the oven to 350°F and line a baking sheet with parchment paper.

To make the dukkah, add all ingredients except the salmon to a food processor or blender. Pulse until walnut pieces are small and crumbly.

Heat a small or medium-size frying pan over medium heat. Toast the dukkah for 3–5 minutes or until fragrantly nutty. Remove from frying pan and transfer to a heat-safe dish if not using immediately.

Place salmon skin side down on the prepared baking sheet. Evenly distribute the walnut dukkah on top of the salmon fillets, gently pressing it onto the flesh of the salmon so it sticks.

Bake for 15–20 minutes or until the fish flakes easily when pulled with a fork.

Serve with Lemony Kale (see next page).

# Lemony Kale

**SERVES 4**

- 2 tablespoons extra-virgin olive oil
- 4–5 garlic cloves, minced
- 2 large bunches kale, woody stems removed, roughly chopped
- 1/4 teaspoon black pepper
- 1/4 teaspoon sea salt
- 2 tablespoons lemon juice

In a large frying pan or large sauce pot, heat oil over medium heat. Add the garlic and sauté for 1 minute.

Add kale, pepper, sea salt, and lemon juice and sauté until kale is wilted and bright green, about 3–4 minutes. Adjust seasoning with salt, pepper, and additional lemon juice if desired. Serve immediately.

Contributed by NeuroTrition Inc.

# Burrito Bowl with Yogurt Cilantro Dressing

| Healthy Food Index | Calories | MACs | Polyphenols | Omega-3 FA/ Total Fats |
|---|---|---|---|---|
| 3.02 | 500 | 16.4 | 0.167 | 0.03 |

(listed amount per serving)

**SERVES 4**

FOR THE BURRITO BOWL:

1 cup brown rice
2 tablespoons avocado oil
1 white onion, diced small
4 garlic cloves, minced
1 tablespoon cumin powder
1/4–1/2 teaspoon chili flakes
28-ounce can diced tomatoes
1 cup organic frozen corn
2 cups cooked black beans
Juice of 1 lime
1 teaspoon sea salt
Pepper to taste
4 cups bitter greens, such as arugula, spinach, or massaged kale

FOR THE YOGURT CILANTRO DRESSING:

7 ounces coconut yogurt
1/2 bunch cilantro or parsley
Pinch of sea salt
Fresh jalapeño *(optional)*

FOR GARNISH:

1 jalapeño, sliced thin
2 avocados, sliced

Cook brown rice as directed on the package (usually takes 35 minutes).

In a large frying pan, heat avocado oil over medium heat. Add onion and cook for 5 minutes, then add garlic, cumin, and chili flakes. Cook for 3 minutes.

Add canned tomatoes, corn, black beans, lime juice, salt, and pepper. Cook until most of the tomato juice has evaporated (5–10 minutes).

While the bean mixture is cooking, make the Yogurt Cilantro Dressing:

Mix the coconut yogurt, cilantro, and a pinch of salt (as well as jalapeño if using) in a blender and blend on high speed until smooth.

To serve, portion rice in a bowl with bean-and-corn medley along with fresh bitter greens, garnish with sliced avocado and jalapeño, and drizzle with the dressing.

Developed by NeuroTrition Inc.

# Spinach and Cheese Frittata

| Healthy Food Index | Calories | MACs | Polyphenols | Omega-3 FA/ Total Fats |
|---|---|---|---|---|
| 1.6 | 310 | 4.3 | 0.03 | 0.03 |

(listed amount per serving)

**SERVES 4**

3 tablespoons extra-virgin olive oil
1 leek, white and pale green parts only, halved lengthwise, rinsed, and thinly sliced crosswise
1 medium onion, thinly sliced
2 cups tomatoes, sliced
3/4 teaspoon sea salt
Freshly ground pepper
1/2 teaspoon turmeric
1 tablespoon grated fresh ginger
2 cups lightly packed baby spinach, coarsely chopped
4 large eggs, beaten
3 ounces goat cheese

FOR GARNISH:
1 avocado, sliced thinly
1/4 cup cut fresh dill
3 teaspoons chopped chives
Zest of 1/2 lemon

In a 10-inch skillet, heat the oil over medium heat.

Add the leek and onion and sauté until soft and translucent, about 5 minutes.

Stir in tomatoes, salt, pepper, turmeric, and ginger, and cook another 4 minutes.

Add the spinach and stir until the leaves are wilted.

Add the egg mixture, and after about a minute, turn heat to medium-low.

Add the goat cheese to the frittata and let it cook until the eggs have set, about 5–6 minutes.

Transfer frittata to a plate and top with sliced avocados, dill, chives, and lemon zest.

Contributed by Minou Mayer, MA

# Breaded Chicken with Vegetables

| Healthy Food Index | Calories | MACs | Polyphenols | Omega-3 FA/ Total Fats |
|---|---|---|---|---|
| 1.80 | 299 | 5 | 0.012 | 0.039 |

(listed amount per serving)

**SERVES 4**

2 tablespoons extra-virgin olive oil
Salt and pepper
1/2 teaspoon cloves
3/4 cup bread crumbs
4 organic, free-range chicken cutlets (thin chicken breast)
1 onion, sliced
2 shallots, sliced
4 garlic cloves, minced
3 cups broccoli florets
3 cups cauliflower florets
1/2 cup chicken broth or water
2 teaspoons dry oregano
2 teaspoons fresh thyme
Juice of 1 lemon
Fresh chopped cilantro for garnish

Heat olive oil in a skillet on medium-high heat. Make sure the oil is hot before frying chicken.

Add about 1/2 teaspoon salt, 1/2 teaspoon ground pepper, and cloves to the bread crumbs and mix well.

Pat each chicken cutlet into the bread crumbs mix and fry in the skillet about 2 minutes on each side until they have a nice golden color.

Remove chicken from skillet and place on a plate.

In a skillet, add a little olive oil on medium heat.

Sauté onion and shallots for about 3–4 minutes.

Add garlic into the skillet and sauté for a minute.

Add broccoli and cauliflower florets and chicken broth or water.

Sprinkle with salt and pepper.

Place chicken cutlets onto the vegetable mix.

Add oregano and thyme to the lemon juice and pour over the chicken.

Cook 5–7 minutes, until the chicken temperature is about 160°F.

Add cilantro for garnish.

Contributed by Minou Mayer, MA

# Cauliflower Chickpea Couscous

| Healthy Food Index | Calories | MACs | Polyphenols | Omega-3 FA/ Total Fats |
|---|---|---|---|---|
| 1.95 | 487 | 10.5 | 0.06 | 0.02 |

(listed amount per serving)

**SERVES 4**

3 tablespoons extra-virgin olive oil
1 medium yellow onion, finely chopped
1 each yellow and orange bell pepper, seeded, stemmed, and cut into medium squares
3 cups cauliflower florets
1/2 teaspoon each cumin powder, turmeric powder, and dried thyme
1 tablespoon fresh ground ginger
2 teaspoons cinnamon powder (or a pinch or two of your favorite hot red pepper flakes)
2 cups fresh tomato slices (or mild salsa)
1/2 cup water or broth
1 can organic chickpeas
3 tablespoons lemon juice
1/2 cup cilantro or parsley
1 1/2 cups uncooked couscous
Salt and pepper

Heat 2 tablespoons oil in a large pot over medium heat.

Add onion and stir occasionally until the onion is golden and soft, about 5 minutes.

Add peppers, cauliflower, spices, and fresh tomatoes, and sauté about 3 minutes.

Add 1/2 cup water or broth.

Cook over medium heat for about 7–8 minutes.

Add chickpeas, with liquid.

Lower heat and simmer about 5 minutes more until cauliflower is tender but still slightly crisp.

Add lemon juice and most of cilantro or parsley (keeping the rest for garnish).

While the chickpea mixture is simmering:

Put couscous in a medium heatproof bowl.

Stir in 1 tablespoon oil, 1/2 teaspoon kosher salt, and several grinds of black pepper.

Gradually stir in 1½ cups boiling water.

Cover and let it sit 8–10 minutes.

Fluff with a fork.

Spoon couscous into bowls, top with the vegetables, and garnish with cilantro/parsley.

Contributed by Minou Mayer, MA

# Saag Paneer

| Healthy Food Index | Calories | MACs | Polyphenols | Omega-3 FA/ Total Fats |
|---|---|---|---|---|
| 1.04 | 550 | 3.3 | 0.1 | 0.05 |

(listed amount per serving)

**SERVES 4**

- 1 pound spinach, chopped
- 2 teaspoons dried fenugreek
- 4 tablespoons ghee (if not available, use EVOO)
- 12 ounces paneer, cut into 1/2-inch cubes
- 1 yellow onion, chopped
- 2 cloves garlic, minced
- 1 teaspoon fresh ginger, minced
- 2 teaspoons cumin
- 2 teaspoons garam masala
- 1/2 teaspoon turmeric
- 1/4 teaspoon cayenne pepper
- 1/4 teaspoon sea salt
- 1 1/2 cups coconut milk

Add the spinach and fenugreek to boiling water and cook 2–3 minutes.

Drain well, squeezing out as much liquid as possible before chopping the spinach finely.

Add the ghee to a pan and fry the paneer cubes until lightly browned, then remove from pan.

Add the onion, garlic, and ginger to the ghee and cook, stirring, on medium heat until wilted and translucent.

Add spinach, cumin, garam masala, turmeric, cayenne pepper, sea salt, and coconut milk (or cream), along with the browned paneer.

Cook uncovered 10–15 minutes or until the coconut milk/cream has cooked down, resulting in a thick green spinach sauce.

Contributed by Arpana Gupta, PhD

## Super Bowls

### Power Bowl

| Healthy Food Index | Calories | MACs | Polyphenols | Omega-3 FA/ Total Fats |
|---|---|---|---|---|
| 3.8 | 199 | 5.43 | 0.39 | 0.31 |

(listed amount per serving)

**SERVES 1**

2 tablespoons steel-cut oats
1 teaspoon flaxseed
1 teaspoon unsalted roasted sunflower seeds
1 teaspoon chia seeds
1 teaspoon raw hempseed
1 teaspoon pumpkin seeds
1/4 cup unfiltered apple juice
1/2 cup nondairy milk
2 ounces seasonal berries (blueberries, strawberries, raspberries, blackberries)

In a medium-size bowl, mix the oats and seeds. Pour in unfiltered apple juice and nondairy milk and stir. Top with berries and enjoy.

Contributed by Minou Mayer, MA

# Tropical Bowl

| Healthy Food Index | Calories | MACs | Polyphenols | Omega-3 FA/ Total Fats |
|---|---|---|---|---|
| 3.34 | 300.8 | 6.9 | 1.28 | 0.42 |

(listed amount per serving)

**SERVES 1**

1 teaspoon açai berry powder
1 teaspoon dried goji berries
1 teaspoon fresh or dried Inka berries (Peruvian groundcherry)
1 fresh date, cut into small pieces
1 teaspoon chia seeds
1 teaspoon cacao powder
1 teaspoon raw hempseed
1/2 cup unsweetened nondairy milk or unsweetened fermented milk
1/4 cup unfiltered, unsweetened apple juice
1 fresh fig, sliced
1 slice fresh pineapple, cut into small pieces
1 slice fresh mango, cut into small pieces

In a medium-size bowl, add the first seven ingredients. Pour in nondairy milk and unfiltered apple juice and stir. Top with fruit and enjoy.

Contributed by Minou Mayer, MA

## Fiber Bowl

| Healthy Food Index | Calories | MACs | Polyphenols | Omega-3 FA/ Total Fats |
|---|---|---|---|---|
| 4.0 | 330.75 | 9.13 | 0.75 | 0.47 |

(listed amount per serving)

**SERVES 1**

- 1 piece canned jackfruit, cut into small pieces
- 1 tablespoon chia seeds
- 1 tablespoon ancient grain flakes (I use Nature's Path Heritage Flakes)
- 1 tablespoon oat bran
- 1 tablespoon raw hempseed
- 1/2 cup kefir or unsweetened nondairy fermented yogurt
- 1/4 cup unfiltered, unsweetened apple juice
- 1/2 apple, cut into small pieces
- 2 prunes, cut into small pieces

In a medium-size bowl, stir together jackfruit with chia, grain flakes, oat bran, and hempseed. Add kefir or yogurt and unfiltered apple juice and stir. Top with fruit and enjoy.

Contributed by Minou Mayer, MA

# Polyphenol Bowl

| Healthy Food Index | Calories | MACs | Polyphenols | Omega-3 FA/ Total Fats |
|---|---|---|---|---|
| 3.9 | 208.9 | 5.79 | 0.32 | 0.28 |

(listed amount per serving)

**SERVES 1**

1 teaspoon dried goji berries
1 teaspoon chia seeds
1 teaspoon roasted pumpkin seeds
1 teaspoon roasted sunflower seeds
1 tablespoon nuts (hazelnuts, pecans, or walnuts)
1 tablespoon oat bran
1 teaspoon cacao powder
1 teaspoon maqui powder *(optional)*
1 teaspoon camu camu powder *(optional)*
1/2 cup hemp milk or unsweetened nondairy yogurt
1/4 cup unfiltered, unsweetened apple juice
1 tablespoon seasonal fruit (organic blueberries, strawberries, raspberries, blackberries, plums, pomegranates, Inka berries)

In a medium-size bowl, add the first nine ingredients. Pour in nondairy milk or yogurt and unfiltered apple juice and stir. Top with fruit and enjoy.

Contributed by Minou Mayer, MA

# Chia Oat Bowl with Fruit

| Healthy Food Index | Calories | MACs | Polyphenols | Omega-3 FA/ Total Fats |
|---|---|---|---|---|
| 5.15 | 414 | 13.9 | 0.694 | 0.26 |

(listed amount per serving)

**SERVES 1**

**2 tablespoons chia seeds**
**2 tablespoons steel-cut oats**
**1 teaspoon vanilla extract**
**1 cup nondairy milk**
**1/2 cup of your favorite fruit (apples, bananas, peaches), chopped into small pieces**
**1/4 cup walnuts, chopped**
**1/2 teaspoon cinnamon**
**1 tablespoon cacao *(optional)***

In a blender, mix chia seeds, oats, vanilla, and plant-based milk.

Pour mixture into a bowl, cover, and refrigerate for a few hours or overnight.

Top with fruit, walnuts, cinnamon, and cacao (if desired) and enjoy.

Contributed by Minou Mayer, MA

# Seed Parfait

| Healthy Food Index | Calories | MACs | Polyphenols | Omega-3 FA/ Total Fats |
|---|---|---|---|---|
| 3.98 | 542 | 18.8 | 0.591 | 0.15 |

(listed amount per serving)

**SERVES 1**

- 1 cup unsweetened plant-based yogurt
- 1 tablespoon chia seeds
- 1 tablespoon flaxseed
- 1/2 cup toasted oats
- 2 tablespoons crushed toasted almonds
- 1/2 cup blueberries
- 1/2 teaspoon cacao
- 1/2 teaspoon cinnamon

In a medium-size bowl, mix yogurt with chia seeds and flaxseed.

Add toasted oats and almonds.

Add blueberries and top it off with cacao and cinnamon.

Contributed by Minou Mayer, MA

## Smoothies

# Pomegranate Chocolate Smoothie

| Healthy Food Index | Calories | MACs | Polyphenols | Omega-3 FA/ Total Fats |
|---|---|---|---|---|
| 3.8 | 292 | 12.1 | 0.61 | 0.09 |

(listed amount per serving)

**SERVES 2**

**8 ounces unsweetened almond milk**
**4 ounces pomegranate juice**
**6 ounces organic baby spinach**
**1 ripe banana, frozen**
**2–3 dates to taste**
**3 tablespoons cacao powder**
**2 cups frozen blueberries**
**1 tablespoon ground flaxseed *(optional)***

Place all ingredients in a high-power blender and process until smooth.

Contributed by Chef AJ

# Mango Ginger Tango

| Healthy Food Index | Calories | MACs | Polyphenols | Omega-3 FA/ Total Fats |
|---|---|---|---|---|
| 1.4 | 370 | 6 | 0.256 | 0.4 |

(listed amount per serving)

**SERVES 1–2**

**1 cup frozen mango cubes**
**1½ cups unfiltered organic apple juice**
**1 banana**
**½ tablespoon ground ginger**
**1 teaspoon ground cinnamon**

In a blender combine all ingredients until smooth.

Contributed by Minou Mayer, MA

# Green Machine

| Healthy Food Index | Calories | MACs | Polyphenols | Omega-3 FA/ Total Fats |
|---|---|---|---|---|
| 3.4 | 255 | 7.75 | 0.01 | 0.06 |

(listed amount per serving)

**SERVES 1–2**

1 cup flax milk or hemp milk
1/2 cup frozen spinach
1/2 avocado
1/2 cup cilantro
1 teaspoon ground ginger
1/2 teaspoon ground cloves
1/2 teaspoon ground black pepper
Mint leaves *(optional)*

In a blender combine all ingredients and blend until smooth.

Contributed by Minou Mayer, MA

# Berry Fantasy

| Healthy Food Index | Calories | MACs | Polyphenols | Omega-3 FA/ Total Fats |
|---|---|---|---|---|
| 4.0 | 264 | 11.3 | 0.49 | 0.09 |

(listed amount per serving)

**SERVES 2**

1 banana
1 cup frozen strawberries
1 cup frozen blueberries
1 cup frozen raspberries
1/2 cup nondairy yogurt
2 1/2 cups flax milk or hemp milk
1 teaspoon ground cinnamon

In a blender combine all ingredients and blend until smooth. If the mixture is too thick, you can add more milk and blend.

Contributed by Minou Mayer, MA

## Salads

# Ancient Grain Salad

| Healthy Food Index | Calories | MACs | Polyphenols | Omega-3 FA/ Total Fats |
|---|---|---|---|---|
| 2.9 | 306 | 10.3 | 0.048 | 0.013 |

(listed amount per serving)

**SERVES 2**

- 1/2 cup whole-grain red bulgur
- 1/2 cup boiling water
- 1 cup canned garbanzo beans
- 1/2 cup scallions, sliced thinly
- 1 cup tomatoes, cut into small pieces
- 2 tablespoons lemon juice
- Salt and pepper
- 1 teaspoon Italian seasoning
- 1/4 cup parsley, chopped finely

Combine bulgur and boiling water and soak for about an hour.

Drain well and squeeze out any excess water.

In a large bowl, add garbanzo beans to the bulgur.

Add scallions, tomatoes, lemon juice, salt, pepper, spices, and parsley to the mixture and mix well.

Contributed by Minou Mayer, MA

# Braised Red Cabbage

| Healthy Food Index | Calories | MACs | Polyphenols | Omega-3 FA/ Total Fats |
|---|---|---|---|---|
| 3.1 | 210 | 7 | 0.176 | 0.02 |

(listed amount per serving)

**SERVES 4**

- 2 tablespoons extra-virgin olive oil
- 1 large yellow onion, finely sliced
- 2–3 tart apples, such as Granny Smith, cored and peeled and sliced
- 1 large red cabbage, cored, quartered, and thinly sliced
- 1/2 cup water or chicken broth
- 1/2 cup red wine
- Salt and pepper
- 1/2 teaspoon cloves
- 1 teaspoon thyme

In a pot add olive oil over medium heat.

Add onion and sauté for a few minutes until tender.

Add apples and continue to sauté for another couple of minutes

Add cabbage and water (or chicken broth) and bring to a boil over medium heat.

Add red wine and season with salt, pepper, cloves, and thyme.

Stir, reduce heat to medium-low, and cover.

Cook and stir often until cabbage is tender, 30–40 minutes.

Contributed by Minou Mayer, MA

# Spinach and Broccoli Salad with Sauerkraut Dressing

| Healthy Food Index | Calories | MACs | Polyphenols | Omega-3 FA/ Total Fats |
|---|---|---|---|---|
| 3.6 | 399 | 13 | 0.456 | 0.04 |

(listed amount per serving)

**SERVES 2**

FOR THE SALAD:
2 cups baby spinach
1/2 cup cherry tomatoes, cut in half
1 avocado, peeled and cut into small slices or squares
2 tablespoons feta cheese
1/2 cup shelled edamame
Handful of chopped cilantro
2 cups steamed broccoli florets

FOR THE DRESSING:
2 tablespoons extra-virgin olive oil
2 tablespoons soy sauce
1/2 cup sauerkraut with juice
Fresh ground pepper

Add all salad ingredients to a large bowl.

In a small bowl, whisk together all dressing ingredients.

Pour dressing over salad and toss to coat. Serve immediately.

Contributed by Minou Mayer, MA

# Garlicky Kale Caesar Salad with Flaxseed Croutons

| Healthy Food Index | Calories | MACs | Polyphenols | Omega-3 FA/ Total Fats |
|---|---|---|---|---|
| 2.0 | 428 | 5.85 | 0.21 | 0.08 |

(listed amount per serving)

**MAKES 4–6 SERVINGS**

FOR THE SALAD:
2 big bunches of kale, stems removed, chopped into bite-size pieces
3 tablespoons cold-pressed flaxseed oil
1/4 teaspoon sea salt

FOR THE DRESSING:
4 garlic cloves, minced (approximately 1 tablespoon)
1 teaspoon anchovy paste
1 teaspoon capers
1 teaspoon caper brine
1/4 teaspoon black pepper
1/4 cup lemon juice, divided in halves
2 free-range egg yolks
1/2 teaspoon sea salt
1 teaspoon mustard powder
1/4 cup extra-virgin olive oil
1/3 cup avocado oil

Place all dressing ingredients, except the olive and avocado oils and half the lemon juice, into a food processor. Process on low-medium until a paste forms (approximately 30–60 seconds).

With the food processor on low speed, very slowly drizzle both oils into the mixture until the oil-and-egg-yolk mixture emulsifies, creating a creamy dressing. Add remaining lemon juice and adjust seasoning with more salt or pepper if necessary.

Next, combine chopped kale, flaxseed oil, and sea salt into a large mixing bowl.

Using clean hands, massage oil and salt into kale until kale begins to become tender.

Combine the marinated kale with desired amount of Caesar dressing. To boost the fiber, polyphenol, and omega-3 content (while also adding some nice crunch), top with Flaxseed Croutons (see next page).

## Flaxseed Croutons

1 cup gold or brown flaxseed meal
1/4 cup coconut flour
1 teaspoon baking soda
1/2 teaspoon sea salt
1/4 teaspoon dried thyme
3 free-range eggs
1/2 cup water
4 garlic cloves, minced (approximately 1 tablespoon)
1/4 cup + 2 tablespoons extra-virgin olive oil (divided)

Preheat the oven to 350°F and line a baking sheet with parchment paper.

In a medium-size bowl, mix flaxseed meal, coconut flour, baking soda, sea salt, and thyme. In a separate bowl, whisk together eggs, water, garlic, and 1/4 cup of the olive oil.

Combine wet ingredients with dry and mix well. Let mixture sit and thicken for 5 minutes.

Transfer mixture onto prepared baking sheet and spread into a 1/2-inch-thick rectangle shape. There is no need to spread it to cover the entire baking sheet. Bake for 20 minutes or until slightly firm. Let cool before cutting into 1-inch cubes.

Return the oven to 350°F. Place croutons on a baking sheet and drizzle with 2 tablespoons olive oil. Toast for 10–15 minutes, depending on how crunchy you want them.

Developed by NeuroTrition Inc.

# Neuro-Niçoise Salad

| Healthy Food Index | Calories | MACs | Polyphenols | Omega-3 FA/ Total Fats |
|---|---|---|---|---|
| 2.0 | 469 | 4.3 | 0.260 | 0.13 |

(listed amount per serving)

**SERVES 4**

FOR THE SALAD:

1 medium sweet potato, skin on, sliced into 1/4-inch to 1/2-inch–thick rounds
1 tablespoon avocado oil
1/4 teaspoon sea salt
2 cups (about 1/2 pound) green beans, blanched
1 cup cherry tomatoes, sliced in half
1/2 cup Niçoise olives, pitted and sliced in half
4 cups arugula
2 cans of sardines, drained
4 free-range eggs, medium- or hard-boiled, sliced in half

FOR THE DRESSING:

1/2 tablespoon grainy mustard
1 1/2 tablespoons raw apple cider vinegar
2 tablespoons lemon juice
1 garlic clove, minced
1/4 cup parsley, stems removed, leaves finely chopped
1/4 teaspoon sea salt
1/4 teaspoon black pepper
1/4 cup extra-virgin olive oil
2 tablespoons cold-pressed flaxseed oil

Preheat the oven to 350°F and line a baking sheet with parchment paper. In a medium-size bowl, toss together sweet potato, avocado oil, and salt, then transfer to the prepared baking sheet. Bake 15–20 minutes or until the sweet potato is tender when poked with a fork.

While the potato is roasting, make the salad dressing. Add the mustard, vinegar, lemon juice, garlic, parsley, sea salt, and pepper to a medium-size bowl and whisk together. Continue to whisk as you slowly drizzle in olive and flaxseed oils until all the dressing ingredients are combined.

In a large bowl, combine potato, green beans, tomatoes, olives, and arugula, then toss with desired amount of dressing. Divide the salad evenly among 4 bowls, then top each serving with 1/4 of the sardines and 2 halves of a boiled egg. Garnish with black pepper if desired.

Developed by NeuroTrition Inc.

# Butter Lettuce, Avocado, and Citrus Salad

| Healthy Food Index | Calories | MACs | Polyphenols | Omega-3 FA/ Total Fats |
|---|---|---|---|---|
| 4.0 | 253 | 8.92 | 0.18 | 0.14 |

(listed amount per serving)

**SERVES 2–3**

FOR THE SALAD:
8 ounces crisp butter lettuce, torn
1 Persian cucumber, sliced thinly
1 cup tomatoes, sliced
2 oranges, peeled and cut into small pieces
2 Fuji apples, cored, sliced, and cut into small pieces
1 avocado, peeled and pitted, cut into squares
1/2 cup toasted sunflower seeds

FOR THE DRESSING:
1 tablespoon extra-virgin olive oil
2 tablespoons orange juice
1 tablespoon soy sauce

Mix lettuce, cucumber, tomatoes, oranges, and apples.

Add avocado pieces and sprinkle with toasted sunflower seeds.

Whisk together the olive oil, orange juice, and soy sauce to make dressing.

Pour dressing over salad, toss to coat, and serve.

Contributed by Minou Mayer, MA

# Beet Salad with Goat Cheese

| Healthy Food Index | Calories | MACs | Polyphenols | Omega-3 FA/ Total Fats |
|---|---|---|---|---|
| 3.2 | 278 | 6.18 | 0.071 | 0.17 |

(listed amount per serving)

**SERVES 2**

2 medium red or golden beets, washed and green parts removed
3 cups baby spinach
2 oranges, peeled and segmented
2 teaspoons fresh chives, chopped
2 teaspoons fresh thyme, chopped
1 tablespoon extra-virgin olive oil
1 tablespoon balsamic vinegar
1/2 cup toasted walnuts
1/2 cup crumbled goat cheese
Salt and pepper to taste

Cook beets until tender, about 20 minutes. Peel and slice when cool.

Place spinach in a medium serving bowl. Add oranges and beets.

In a small bowl, whisk together chives, thyme, olive oil, and vinegar.

Pour dressing over salad.

Top with toasted walnuts and goat cheese. Add salt and pepper to taste, and serve.

Contributed by Minou Mayer, MA

# Mung Bean Sprout Salad

| Healthy Food Index | Calories | MACs | Polyphenols | Omega-3 FA/ Total Fats |
|---|---|---|---|---|
| 5.9 | 157 | 9.75 | 0.53 | 0.75 |

(listed amount per serving)

**SERVES 4**

- 2 cups sprouted mung beans
- 1 small or medium onion, finely chopped
- 1 medium tomato, finely chopped
- 1 green chili *(optional)*, finely chopped
- 1/4 teaspoon red chili powder
- 1/2 teaspoon chaat masala *(optional)*
- 1 boiled potato or sweet potato *(optional)*
- Rock salt or black salt as desired
- 1 teaspoon lemon juice, as desired
- A few coriander leaves and lemon slices for garnish

Wash mung beans thoroughly. Drain and soak in plenty of water for 6 to 8 hours or overnight.

Drain the soaked beans and place in a large bowl, making sure there is some moisture left on them.

Cover the bowl with a lid and keep in a warm place for about 8 to 12 hours (mung beans start sprouting faster in warm weather).

Refrigerate leftover sprouts.

Rinse the sprouted mung beans in water, then steam or boil until completely cooked. Strain.

In a bowl, mix all ingredients except salt and lemon juice.

Season with salt and add a few drops of lemon juice. Garnish with lemon slices and coriander leaves. Serve immediately.

Contributed by Arpana Gupta, PhD

# Avocado Hummus Dip

| Healthy Food Index | Calories | MACs | Polyphenols | Omega-3 FA/ Total Fats |
|---|---|---|---|---|
| 2.6 | 150 | 4 | 0.08 | 0.008 |

(listed amount per serving)

**SERVES 4**

**4 cloves garlic**
**1 teaspoon chili flakes**
**½ teaspoon cumin powder**
**½ cup canned garbanzo beans, drained**
**2 tablespoons lemon juice**
**½ teaspoon turmeric powder**
**1 tablespoon fresh ground ginger**
**1½ avocados, peeled and pitted**
**1 tablespoon extra-virgin olive oil**
**Salt and freshly ground pepper**
**½ teaspoon paprika for garnish**
**1 teaspoon chopped parsley for garnish**

Blend garlic, chili flakes, cumin, garbanzo beans, lemon juice, turmeric, and ginger in a food processor.

Add avocado and blend another 20 seconds.

Put mixture into a bowl, add olive oil, salt, and pepper, and mix. Garnish with paprika and parsley.

Contributed by Minou Mayer, MA

# Kale and Lentil Salad

| Healthy Food Index | Calories | MACs | Polyphenols | Omega-3 FA/ Total Fats |
|---|---|---|---|---|
| 3.3 | 307 | 7.8 | 0.027 | 0.1 |

(listed amount per serving)

**SERVES 2**

FOR THE SALAD:

- 3/4 cup green lentils
- 1 large head of Tuscan kale, stems removed and discarded, leaves finely chopped
- 1 cup cherry tomatoes, cut into halves
- 1 avocado, cut into small slices or squares
- Handful of chopped cilantro
- 1/2 cup chopped walnuts, lightly toasted
- Salt and freshly ground pepper

FOR THE DRESSING:

- 2 tablespoons extra-virgin olive oil
- 1 tablespoon fresh lemon juice
- ½ cup sauerkraut
- 1 teaspoon cumin powder
- 1/2 teaspoon fresh ground pepper

Add lentils to a large pot of salted boiling water and cook 20–25 minutes, until tender. Drain and let cool.

Add kale, tomatoes, lentils, avocado, and cilantro to a large bowl.

In a small bowl, whisk together all dressing ingredients.

Pour dressing over salad and top with toasted walnuts and salt and pepper to taste.

Contributed by Minou Mayer, MA

# Sautéed Vegetables with Mustard Vinaigrette

| Healthy Food Index | Calories | MACs | Polyphenols | Omega-3 FA/ Total Fats |
|---|---|---|---|---|
| 3.0 | 282 | 8.2 | 0.07 | 0.03 |

(listed amount per serving)

**SERVES 4**

3 tablespoons extra-virgin olive oil
1 red onion, cut in half, then into 1-inch slices
2 cups carrots, peeled and sliced diagonally
½ tablespoon fresh ground ginger
4 cloves garlic, peeled and sliced
1 cup zucchini, sliced or cut into 1-inch squares
2 cups different-colored bell pepper, cut into 1-inch squares
2 cups broccoli florets
2 cups cauliflower florets
Salt and pepper to taste
1 cup fresh cherry tomatoes, cut in half
1 cup chickpeas, drained

FOR THE DRESSING:
Handful of chopped parsley
2 tablespoons mustard
1 tablespoon extra-virgin olive oil
2 tablespoons wine vinegar
1 teaspoon thyme

Put all dressing ingredients in a jar and shake well.

In a large frying pan, heat the 3 tablespoons of olive oil on medium heat.

Sauté onion and carrots for about 3 minutes.

Add ginger, garlic, zucchini, bell pepper, broccoli, and cauliflower to the pan.

Sprinkle salt and pepper.

Cover and cook 3–4 minutes, until tender.

Stir and add ¼ cup water to the pan and cook for a couple more minutes.

Make sure you don't overcook the vegetables, as you want them crisp.

Let vegetables cool to room temperature and place in a large bowl.

Add tomatoes and chickpeas.

Pour dressing over the mixture, toss, and serve.

Contributed by Minou Mayer, MA

## Sandwiches

# Avocado Egg Toast

| Healthy Food Index | Calories | MACs | Polyphenols | Omega-3 FA/ Total Fats |
|---|---|---|---|---|
| 1.4 | 331 | 4.8 | 0.153 | 0.005 |

(listed amount per serving)

**SERVES 2**

1 tablespoon extra-virgin olive oil
½ teaspoon turmeric
2 eggs
Salt and pepper
2 slices wheat sourdough bread
½ ripe avocado
1 small red onion, thinly sliced
½ tomato, thinly sliced

Heat oil in a frying pan on medium-low heat.

Add turmeric to the oil and let it sizzle for a second.

Crack the eggs into the pan and season with salt and pepper.

Cover and cook the eggs, however you like them, about 3–4 minutes.

Toast slices of bread and mash avocado on them.

Arrange the onion and tomato slices on bread and top with an egg.

Contributed by Minou Mayer, MA

# Spanish Mackerel Salad Sandwiches

| Healthy Food Index | Calories | MACs | Polyphenols | Omega-3 FA/ Total Fats |
|---|---|---|---|---|
| 1.2 | 450 | 5.15 | 0.028 | 0.1 |

(listed amount per serving)

**SERVES 2**

2 tins roasted Atlantic garlic mackerel fillets (such as Patagonia Provisions), drained and separated into fat flakes
1 celery stalk, diced small
1 small shallot or 1/4 red onion, diced small
About 1/3 cup roughly chopped flat-leaf parsley leaves
1 tablespoon lemon juice
1 teaspoon lemon zest
1 teaspoon Dijon mustard
Salt and pepper

FOR THE SANDWICHES:
Slices of Muenster or Swiss cheese (optional)
4 slices whole-wheat bread
6–8 dill pickle chips, drained and patted dry *(optional)*

In a medium bowl, gently mix mackerel salad ingredients together.

Heat a nonstick pan over medium heat.

Assemble sandwiches: Lay a cheese slice on a slice of bread. Add 3–4 pickle chips, half of the mackerel salad, and another cheese slice; top with bread. Repeat for the second sandwich.

Place sandwiches in pan.

Cook until golden brown and cheese is melted, about 5 minutes per side.

**Variations:**

- Add a dash of curry powder, dukkah, harissa, or any other savory spice mix.
- Stir in chopped fresh herbs.
- If you don't want added calories and animal fat, you may skip the cheese.

Modified from Patagonia Provisions, Inc.

# Veggie Burger

| Healthy Food Index | Calories | MACs | Polyphenols | Omega-3 FA/ Total Fats |
|---|---|---|---|---|
| 2.2 | 413 | 11 | 0.001 | 0.003 |

(listed amount per serving)

**SERVES 4**

- **5.8-ounce pouch Patagonia Provisions black-bean soup**
- **1 cup bread crumbs**
- **1/2 cup pepitas (pumpkin seeds), chopped fine or pulsed in food processor**
- **1/4 cup green onions, sliced**
- **1 egg, beaten**
- **1 teaspoon lemon juice**

Cook the black bean soup with half the amount of water (1 cup) and cool. You should have a very thick bean paste.

In a medium mixing bowl, combine remaining ingredients and mix thoroughly with cooled bean mixture.

Divide mixture into four patties and cook on stovetop or grill.

**Stovetop:**

To a hot frying pan, add 2 tablespoons of oil and sear patties until crispy and heated throughout.

**Grill:**

Place patties in a single layer on a lined tray or plate and freeze until firm, 20–30 minutes. Preheat the grill and wipe clean grates with an oiled cloth. Grill each patty 5–7 minutes on each side.

Top with avocado, sprouts, and your favorite condiments.

Adapted from Patagonia Provisions, Inc.

## Desserts

# B-RAW-nie

| Healthy Food Index | Calories | MACs | Polyphenols | Omega-3 FA/ Total Fats |
|---|---|---|---|---|
| 3.3 | 340 | 7 | 0.16 | 0.27 |

(listed amount per serving)

**SERVES 8**

2 cups walnuts
1/2 cup cocoa powder
2 cups pitted dates
1 tablespoon alcohol-free vanilla extract

In a food processor fitted with the S blade, process walnuts to a powder. Do not overprocess to a nut butter.

Add cocoa and process again.

Add dates and process until a ball forms.

Add vanilla and briefly process again.

Transfer the batter to a silicone brownie mold or an 8" × 8" square pan and press in evenly.

Cover and freeze until firm, about 2–3 hours, then slice into squares.

Contributed by Chef AJ

# Cacao Yogurt with Mixed Berries

| Healthy Food Index | Calories | MACs | Polyphenols | Omega-3 FA/ Total Fats |
|---|---|---|---|---|
| 4.13 | 160 | 6.2 | 0.44 | 0.14 |

(listed amount per serving)

**SERVES 1**

**1/2 cup plain plant-based yogurt alternative**
**1 tablespoon cacao**
**1 cup seasonal berries, cut into bite-size pieces if necessary**

Mix cacao and yogurt in a small bowl and stir until smooth.

Top with fruit and enjoy.

Contributed by Minou Mayer, MA

# Pressure-Cooker Blueberry-Millet Pudding

| Healthy Food Index | Calories | MACs | Polyphenols | Omega-3 FA/ Total Fats |
|---|---|---|---|---|
| 1.2 | 256 | 3.12 | 0.49 | 0.13 |

(listed amount per serving)

**SERVES 4–6**

FOR THE PUDDING:
1 cup millet
3 cups unsweetened nondairy milk
1 teaspoon cinnamon
½ teaspoon cardamom
1 teaspoon vanilla powder *(optional)*

FOR THE FRUIT TOPPING:
2 tablespoons date paste
2 cups unsweetened pomegranate juice
4 tablespoons cornstarch dissolved in 4 tablespoons water
1 cup wild blueberries

Place all pudding ingredients in an Instant Pot electric pressure cooker and cook on high for 10 minutes. Release pressure after 10 minutes. This can be enjoyed warm or cold. Millet thickens as it cools.

In a medium saucepan dissolve date paste into the pomegranate juice and reduce to ½ cup of liquid. Slowly stir in cornstarch until mixture thickens, then gently stir in blueberries. Remove from heat.

Distribute pudding mixture evenly into 4–6 tall glasses or parfait dishes. Evenly distribute the fruit topping on each of the parfaits. You can do 2 layers and alternate if you wish.

Chill for a few hours until set.

Contributed by Chef AJ

# World's Healthiest (and Easiest) Pecan Pie

| Healthy Food Index | Calories | MACs | Polyphenols | Omega-3 FA/ Total Fats |
|---|---|---|---|---|
| 2.4 | 366 | 7.3 | 1.83 | 0.015 |

(listed amount per serving)

**MAKES 10–12 SERVINGS**

FOR THE CRUST:
2 cups raw unsalted pecans
2 cups pitted dates
1 teaspoon vanilla powder *(optional but good)*

FOR THE FILLING:
16 ounces pitted dates soaked in 16 ounces water until soft
1 teaspoon vanilla powder *(optional but good)*
12 ounces raw, unsalted pecans (about 3 cups), finely ground to a powder

**First, make the crust:**

Place pecans in a dry food processor fitted with the S blade and process to a flourlike consistency. Do not overprocess, or you will get nut butter.

Add dates and process until a ball forms. You may need to add more dates.

Once the crust is at the proper consistency, add the vanilla and pulse briefly.

Using a piece of parchment paper, press the crust evenly into a 9-inch springform pan.

**Then, make the filling:**

Place dates, soaking liquid, and vanilla into a large food processor fitted with the S blade and puree until smooth.

Add the finely ground pecans and process again until creamy.

**Assemble the pie:**

Remove parchment from crust and pour in filling, spreading evenly.

Decorate the top of the pie with pecan halves.

Freeze the pie overnight or until firm.

Contributed by Chef AJ

## Meal Plans

Here is a guideline for a day's menu. If you plan to follow a time-restricted eating schedule, I recommend two meals a day, lunch and dinner, with healthy snacks in between, within the eight-hour eating period. However, you can also use these recipes in a traditional breakfast, lunch, and dinner meal plan.

- Breakfast: bowl or smoothie
- Brunch/lunch: bowl, sandwich, or salad
- Dinner: simple main (complex mains for weekends only)
- High-fiber/polyphenol snacks with no added sugar in between
    - apples, nuts
    - high-fiber, no-added-sugar bars
    - Navitas Organics snacks
- Drinks
    - black, unsweetened coffee or tea in the morning
    - kombucha
    - green or black tea in the afternoon/evening
    - yerba maté tea
    - unsweetened probiotic drinks or shots
    - water
    - glass of red wine in the evening

# Nutritional Values for Gut-Healthy Food

## High-Fiber Foods

| Ingredient (100g) | Fiber (g/100g) |
|---|---|
| chia seeds | 33.3 |
| cacao | 22.5 |
| flaxseeds | 19.3 |
| lentils | 17.5 |
| oat bran | 16.1 |
| flaxseed meal | 13.3 |
| wheat germ | 12 |
| edamame | 8.8 |
| whole-wheat pasta | 8 |
| pecans | 7.5 |
| black beans | 4.3 |
| wild rice | 4 |
| chickpeas | 2.6 |
| beets | 2.6 |

Source: https://www.nal.usda.gov/sites/www.nal.usda.gov/files/total_dietary_fiber.pdf

## High-Polyphenol Foods

| Ingredient | Total Phenolic Content (mg/100g) |
|---|---|
| chia seeds | 2941.2 (including linolenic polyunsaturated fatty acids) |
| flaxseed | 956.9 (including linolenic polyunsaturated fatty acids and secoisolariciresinol diglucoside) |
| flaxseed oil | 900 (including ferulic acid 4-O-glucoside) |
| coffee | 895 (chlorogenic acid) |
| unpasteurized sauerkraut | 825 (including pinoresinol and kaempferol) |
| blueberries | 310 (including 5-caffeoylquinic acid) |
| cacao powder | 225 (flavanols ) |
| red wine | 220 (including resveratrol and tannin) |
| plum slices | 185 (including 3-caffeoylquinic acid) |
| black beans | 174 (including delphinidin 3-O-glucoside) |
| green tea | 105 (L-theanine) |
| oyster mushrooms | 67 (including ergothioneine) |
| extra-virgin olive oil | 50 (luteolin and oleocanthal) |

Source: http://phenol-explorer.eu/

## High Omega-3 Fatty Acid–Containing Foods

| Ingredient (100g) | Omega-3 FA (mg/100g) |
|---|---|
| flaxseed | 22,800 |
| chia seeds | 18,100 |
| walnuts | 9,200 |
| hempseed | 8,700 |
| flaxseed oil | 8,200 |
| mackerel | 5,100 |
| sardines | 4,000 |
| wild salmon | 2,300 |
| soybeans | 1,400 |
| pecans | 860 |
| tofu | 582 |

Source: https://fdc.nal.usda.gov/

# ACKNOWLEDGMENTS

I am thankful to many individuals who have influenced the progression of my thinking during the past five years about the close interconnectedness between our own health, the health of the environment that produces our food, the health of our plant-based food, and the health of the planet. These influences have convinced me to write a second book going way beyond the basic concepts laid out in *The Mind-Gut Connection*, which was limited to the close interactions between our brain, the gut, and its microbiome. The person who has played the most prominent role in expanding my view from the gut and the brain to the health of the soil and the planet is Yvon Chouinard, the visionary founder of Patagonia, whose life story, philosophy, and passionate fight to save the planet have had a profound and lasting influence on my worldview. And I wouldn't have had the determination to embark on this project without the trust and encouragement of Julie Will, my amazing editor at Harper Wave, who has provided invaluable feedback throughout the entire writing period.

I am thankful to all the patients I have seen in my clinic over the decades whose life stories have taught me about the importance of mind-gut interactions for health and disease, and continue to help me validate the direct clinical relevance of my research findings. Along the same lines, I am incredibly thankful for all the positive feedback I got from readers of *The Mind-Gut Connection,* who often recognize themselves in the patient stories and want more information about how they can reestablish a healthy gut microbiome. My ideas for this book could not have emerged without my close interactions with my research team at UCLA, in particular Dr. Annie Gupta, a major driving force in our studies about the influence of food on gut microbiome–brain interactions, and several bright students working at our center who took a keen interest in contributing to this book's development, in particular Karina Nance and Juliette Frank. I am grateful to be a faculty member in the Division of Digestive Diseases at UCLA, which is at the forefront of research and clinical practice of mind-gut interactions, in particular to my division chief, Dr. Eric Esrailian, who shares many of my views on the interconnection between the mind and the gut. To Drs. Rima Kaddurah-Daouk and Sarkis Mazmanian, who are inspiring scientists leading the effort to unravel the role of the gut microbiome in devastating brain disorders. To Dr. Olaf Sporns, who has been a pioneer in applying advanced network analysis to brain studies, and to Dr. Walter Willet, who has drawn the attention of the scientific world to the close relationship between our health and the climate crisis. To Norbert Niederkofler, the three–Michelin star chef in the Italian Dolomites who combines world-class cuisine with local and sustainable food production, and to Marco Cavalieri, who practices regenerative organic wine and olive oil production in the Adriatic region of Italy.

I am thankful to my coauthor, Nell Casey, who has been tremendously helpful in turning complex scientific concepts into understandable and easily digestible language, and to Clark Miller, who used his artistic creativity to design the illustrations for this book.

And last but not least, I am tremendously thankful to my wife, Minou, and our son, Dylan, with whom I have had ongoing discussions about many details of this book and who have helped with the design and testing of many of the recipes, turning our kitchen into a food laboratory.

# RECIPE RESOURCES

## Contributors

The majority of the recipes in this book were contributed by Orsha Magyar and her Neurochefs from the company NeuroTrition and by my wife, Minou Mayer, MA.

NeuroTrition develops recipes based on nutrition for optimal brain health. You can find out more about this innovative organization at www.neurotrition.ca.

A few recipes were adapted from Patagonia Provisions, Inc. (www.patagoniaprovisions.com).

Dessert recipes were developed by Chef AJ (www.chefajwebsite.com).

Some pasta dishes were adapted from one of my favorite cookbooks, *Jamie's Italy*, by Jamie Oliver (New York: Hyperion, 2006).

## Ingredients

Even though alternative sources are available, based on quality, sustainable production, and health benefits I recommend sourcing ingredients from the following brands:

- Extra-virgin olive oil: Le Corti Dei Farfensi, https://lecortideifarfensiusa.com/collections/olive-oil

- Canned mussels, smoked salmon, canned mackerel, and organic black-bean soup, seeds, and bars: Patagonia Provisions, www.patagoniaprovisions.com
- Canned sardines: wild Portuguese sardines, www.vitalchoice.com
- Fiber bars: NuGo Nutrition, www.nugofiber.com
- Hemp milk and raw hempseed: Manitoba Harvest, www.manitobaharvest.com
- Goji berries, Inka berries, maqui, camu camu, and açai powder: Navitas Organics, www.navitasorganics.com
- Ancient grain flakes: Nature's Path Heritage flakes, www.naturespath.com

# NOTES

## Chapter One: America's Silent Public Health Crisis


1. Eileen M. Crimmins, "Lifespan and Healthspan: Past, Present, and Promise," *Gerontologist* 55, no. 6 (Dec. 2015): 901–11, doi: 10.1093/geront/gnv130, PubMed PMID: 26561272.
2. Centers for Medicare & Medicaid Services, National Health Expenditure Data, Historical, https://www.cms.gov/Research-Statistics-Data-and-Systems/Statistics-Trends-and-Reports/NationalHealthExpendData/NationalHealthAccountsHistorical.
3. Rabah Kamal, Cynthia Cox, and Daniel McDermott, "What Are the Recent and Forecasted Trends in Prescription Drug Spending?" Health System Tracker, Peterson Center on Healthcare and Kaiser Family Foundation, 2019, https://www.healthsystemtracker.org/chart-collection/recent-forecasted-trends-prescription-drug-spending/#item-annual-growth-in-rx-drug-spending-and-total-health-spending-per-capita_nhe-projections-2018-27.
4. Animal Smart, "Comparing Agriculture of the Past with Today," https://animalsmart.org/animals-and-the-environment/comparing-agriculture-of-the-past-with-today.
5. Hilda Razzaghi et al., "10-Year Trends in Noncommunicable Disease Mortality in the Caribbean Region," *Revista Panamericana de Salud Pública*, 2019, 43, doi: 10.26633/RPSP.2019.37.
6. Jean-François Bach, "The Effect of Infections on Susceptibility to Autoimmune and Allergic Diseases," *New England Journal of Medicine* 347, no. 12 (Sept. 19, 2002): 911–20, doi: 10.1056/NEJMra020100.
7. Forough Farrokhyar, E. T. Swarbrink, and E. Jan Irvine, "A Critical Review of Epidemiological Studies in Inflammatory Bowel Disease," *Scandinavian Journal of Gastroenterology* 36, no. 1 (February 2001): 2–15.
8. Nils Åberg et al., "Increase of Asthma, Allergic Rhinitis and Eczema in Swedish Schoolchildren between 1979 and 1991," *Clinical & Experimental Allergy*

25, no. 9 (Sept. 1995): 815–19, doi: 10.1111/j.1365-2222.1995.tb00023.x, PubMed PMID: 8564719.

9. Sigrid Poser et al., "Increasing Incidence of Multiple Sclerosis in South Lower Saxony, Germany," *Neuroepidemiology* 8, no. 4 (1989): 207–13, doi: 10.1159/000110184.
10. H. Okada et al., "The 'Hygiene Hypothesis' for Autoimmune and Allergic Diseases: An Update," *Clinical & Experimental Immunology* 160, no. 1 (Apr. 2010): 1–9, doi: 10.1111/j.1365-2249.2010.04139.x.
11. Michael Ollove, "States Limiting Patient Costs for High-Priced Drugs," Pew Charitable Trusts, July 2, 2015, https://www.pewtrusts.org/en/research-and-analysis/blogs/stateline/2015/07/02/states-limiting-patient-costs-for-high-priced-drugs.
12. Canadian Agency for Drugs and Technologies in Health, "Table 4: Cost-Comparison Table of Biologics for the Treatment of Crohn's Disease," *Common Drug Reviews*, Ottawa, 2017, https://www.ncbi.nlm.nih.gov/books/NBK476194/table/app8.t1.
13. American Autoimmune Related Diseases Association, "Autoimmune Disease List," https://www.aarda.org/diseaselist.
14. Meghan O'Rourke, "What's Wrong with Me?" *New Yorker*, Aug. 26, 2013, https://www.newyorker.com/magazine/2013/08/26/whats-wrong-with-me.
15. Marie Ng et al., "Global, Regional, and National Prevalence of Overweight and Obesity in Children and Adults during 1980–2013: A Systematic Analysis for the Global Burden of Disease Study 2013," *Lancet* 384, no. 9945 (Aug. 30, 2014): 766–81, doi: 10.1016/S0140-6736(14)60460-8, PubMed PMID: 24880830.
16. National Institute of Diabetes and Digestive and Kidney Diseases, "Overweight & Obesity Statistics," https://www.niddk.nih.gov/health-information/health-statistics/overweight-obesity.
17. Mohammad G. Saklayen, "The Global Epidemic of the Metabolic Syndrome," *Current Hypertension Reports* 20, no. 2 (Feb. 2018): 12–20, doi: 10.1007/s11906-018-0812-z. PubMed PMID: 29480368.
18. M. Aguilar et al., "Prevalence of the Metabolic Syndrome in the United States, 2003-2012," *JAMA* 313, no. 9 (2015): 1973–4, doi: 10.1001/jama.2015.4260.
19. American Heart Association, "Cardiovascular Disease: A Costly Burden for America—Projections through 2035," https://healthmetrics.heart.org/wp-content/uploads/2017/10/Cardiovascular-Disease-A-Costly-Burden.pdf.
20. American Heart Association Center for Health Metrics and Evaluation, "Cardiovascular Disease Costs Will Exceed $1 Trillion by 2035, Warns the American Heart Association," press release, Feb. 14, 2017, https://healthmetrics.heart.org/cardiovascular-disease-costs-will-exceed-1-trillion-by-2035-warns-the-american-heart-association.
21. Rebecca Harris et al., "Obesity Is the Most Common Risk Factor for Chronic Liver Disease: Results from a Risk Stratification Pathway Using Transient

Elastography," *American Journal of Gastroenterology* 114, no. 11 (Aug. 2019): 1744–52.

22. Center for Disease Control and Prevention, "Cancers Associated with Overweight and Obesity Make Up 40 Percent of Cancers Diagnosed in the United States," press release, Oct. 3, 2017, https://www.cdc.gov/media/releases/2017/p1003-vs-cancer-obesity.html.
23. Theo Vos et al., "Global, Regional, and National Incidence, Prevalence, and Years Lived with Disability for 328 Diseases and Injuries for 195 Countries, 1990–2016: A Systematic Analysis for the Global Burden of Disease Study 2016," *Lancet* 390, no. 10100 (Sept. 2017): 1211–59, doi: 10.1016/S0140-6736(17)32154-2.
24. Ibid.
25. Parkinson's Foundation, "Statistics," https://www.parkinson.org/Understanding-Parkinsons/Statistics.
26. Autism Speaks, "CDC Increases Estimate of Autism's Prevalence by 15 Percent, to 1 in 59 Children 2018," https://www.autismspeaks.org/science-news/cdc-increases-estimate-autisms-prevalence-15-percent-1-59-children.
27. Mark Rice-Oxley, "Mental Illness: Is There Really a Global Epidemic?" *Guardian*, June 3, 2019.
28. Joseph W. Windsor and Gilaad G. Kaplan, "Evolving Epidemiology of IBD," *Current Gastroenterology Reports* 21, no. 8 (July 2019): 1–9, doi: 10.1007/s11894-019-0705-6.
29. Katarina Zimmer, "There's a Troubling Rise in Colorectal Cancer among Young Adults," *Scientist*, Aug. 26, 2019.
30. American Cancer Society, "Guideline for Colorectal Cancer Screening," https://www.cancer.org/cancer/colon-rectal-cancer/detection-diagnosis-staging/acs-recommendations.html; and Andrew M. D. Wolf et al., "Colorectal Cancer Screening for Average-Risk Adults: 2018 Guideline Update from the American Cancer Society," *CA: A Cancer Journal for Clinicians* 68, no. 4 (July/Aug. 2018): 250–81, doi: 10.3322/caac.21457.
31. Francesco De Virgiliis and Simone Di Giovanni, "Lung Innervation in the Eye of a Cytokine Storm: Neuroimmune Interactions and COVID-19," *Nature Reviews Neurology* 16, no. 11 (Jan. 2020): 645–52, doi: 10.1038/s41582-020-0402-y.
32. Donjete Simnica et al., "The Impact of Western Diet and Nutrients on the Microbiota and Immune Response at Mucosal Interfaces," *Frontiers in Immunology* 8 (July 2017): article 838.

## Chapter Two: A Deeper Connection

1. Johns Hopkins Medicine, "Ayurveda," https://www.hopkinsmedicine.org/health/wellness-and-prevention/ayurveda.
2. René Descartes, *The Method, Meditations and Philosophy of Descartes* (London: Orion, 2004), 15.

3. Wikipedia, "René Descartes," https://en.wikipedia.org/wiki/Ren%C3%A9_Descartes; and Alan Nelson, "Descartes' Dualism and Its Relation to Spinoza's Metaphysics," in: David Cunning, ed., *The Cambridge Companion to Descartes' Meditations* (Cambridge University Press, 2014), 277–98.
4. Wikipedia, "Network Science," https://en.wikipedia.org/wiki/Network_science; and Gosak M et al., "Network science of biological systems at different scales: A review," *Physics of Life Reviews* 24 (2018): 118–35, doi: https://doi.org/10.1016/j.plrev.2017.11.003.
5. C. David Allis and Thomas Jenuwein, "The Molecular Hallmarks of Epigenetic Control," *Nature Reviews Genetics* 17, no. 8 (Aug. 2016): 487–500, doi: 10.1038/nrg.2016.59.
6. Marcus M. Rinschen et al., "Identification of Bioactive Metabolites Using Activity Metabolomics," *Nature Reviews Molecular Cell Biology* 20, no.6 (Feb. 2019): 353–67, doi: 10.1038/s41580-019-0108-4.
7. Maarten Altelaar, Javier Muñoz, and Albert J. R. Heck, "Next-Generation Proteomics: Towards an Integrative View of Proteome Dynamics," *Nature Reviews Genetics* 14, no. 1 (Dec. 2012): 35–48, doi: 10.1038/nrg3356.
8. Pacific Northwest National Laboratory, "Microbiome Science: Confronting Complex Mysteries," https://www.pnnl.gov/microbiome-science.
9. Olaf Sporns, *Discovering the Human Connectome* (Cambridge, MA: MIT Press, 2012).
10. Diego V. Bohórquez and Rodger A. Liddle, "The Gut Connectome: Making Sense of What You Eat," *Journal of Clinical Investigation* 125, no. 3 (Mar. 2015): 888–90, doi: 10.1172/JCI81121.
11. Sporns, *Discovering the Human Connectome*.
12. Clair R. Martin et al., "The Brain-Gut-Microbiome Axis," *Cellular and Molecular Gastroenterology and Hepatology* 6, no. 2 (Apr. 2018): 133–48, doi: 10.1016/j.jcmgh.2018.04.003, PubMed PMID: 30023410.
13. Erica D. Sonnenburg and Justin L. Sonnenburg, "The Ancestral and Industrialized Gut Microbiota and Implications for Human Health," *Nature Reviews Microbiology* 17, no. 6 (June 2019): 383–90, doi: 10.1038/s41579-019-0191-8.
14. Patrice D. Cani, "How Gut Microbes Talk to Organs: The Role of Endocrine and Nervous Routes," *Molecular Metabolism* 5, no. 9 (May 2016): 743–52, doi: 10.1016/j.molmet.2016.05.011, PubMed PMID: 27617197.
15. Siri Carpenter, "That Gut Feeling," *Monitor on Psychology* 43, no. 8 (Sept. 2012): 50.
16. Michael D. Gershon, *The Second Brain: A Groundbreaking New Understanding of Nervous Disorders of the Stomach and Intestine* (New York: Harper Perennial, 1999).
17. Giuseppe Danilo Vighi et al., "Allergy and the Gastrointestinal System," *Clinical & Experimental Immunology* 153, suppl. 1 (Oct. 2008): 3–6, doi: 10.1111/j.1365-2249.2008.03713.x.
18. John B. Furness et al., "The Gut as a Sensory Organ," *Nature Reviews Gas-*

*troenterology & Hepatology* 10, no. 12 (Sept. 2013): 729–40, doi: 10.1038/nrgastro.2013.180.

Chapter Three: The Emerging View of a Healthy Gut Microbiome

1. Abigain Johnson et al., "Daily Sampling Reveals Personalized Diet-Microbiome Associations in Humans," *Cell Host & Microbe* 25, no. 6 (June 2019): 789–802.e5, doi: 10.1016/j.chom.2019.05.005.
2. Jocelyn Kaiser, "There Are About 20,000 Human Genes: So Why Do Scientists Only Study a Small Fraction of Them?" *Science* online, Sept. 18, 2018, https://www.sciencemag.org/news/2018/09/there-are-about-20000-human-genes-so-why-do-scientists-only-study-small-fraction-them.
3. Steve Mao, "Hidden Treasure in the Microbiome," *Science* 365, no. 6458 (Sept. 13, 2019): 1132–33, doi: 10.1126/science.365.6458.1132-g.
4. Mahesh S. Desai et al., "A Dietary Fiber-Deprived Gut Microbiota Degrades the Colonic Mucus Barrier and Enhances Pathogen Susceptibility," *Cell* 167, no. 5 (Nov. 17, 2016): 1339–53.e21, doi: 10.1016/j.cell.2016.10.043.
5. Clinton White House Archives, "President Clinton: Announcing the Completion of the First Survey of the Entire Human Genome 2000," https://clintonwhitehouse3.archives.gov/WH/Work/062600.html.
6. Daniel Aguirre de Cárcer, "The Human Gut Pan-Microbiome Presents a Compositional Core Formed by Discrete Phylogenetic Units," *Scientific Reports* 8, no. 1 (Sept. 2018): article 14069, doi: 10.1038/s41598-018-32221-8, PubMed PMID: 30232462.
7. Catherine A. Lozupone et al., "Diversity, Stability and Resilience of the Human Gut Microbiota," *Nature* 489, no. 7415 (Sept. 13, 2012): 220–30, doi: 10.1038/nature11550.
8. Martin J. Blaser and Stanley Falkow, "What Are the Consequences of the Disappearing Human Microbiota?" *Nature Reviews Microbiology* 7, no. 12 (Nov. 2009): 887–94, doi: 10.1038/nrmicro2245.
9. Christoph A. Thaiss et al., "Transkingdom Control of Microbiota Diurnal Oscillations Promotes Metabolic Homeostasis," *Cell* 159, no. 3 (Oct. 23, 2014): 514–29, doi: 10.1016/j.cell.2014.09.048.
10. Christoph A. Thaiss et al., "Microbiota Diurnal Rhythmicity Programs Host Transcriptome Oscillations," *Cell* 167, no. 6 (Dec. 2016): 1495–1510.e12, doi: 10.1016/j.cell.2016.11.003.
11. Gabriela K. Fragiadakis et al., "Links between Environment, Diet, and the Hunter-Gatherer Microbiome," *Gut Microbes* 10, no. 2 (Aug. 2019): 216–27, doi: 10.1080/19490976.2018.1494103, PubMed PMID: 30118385.
12. Samuel A. Smits et al., "Seasonal Cycling in the Gut Microbiome of the Hadza Hunter-Gatherers of Tanzania," *Science* 357, no. 6353 (Aug. 25, 2017): 802–6, doi: 10.1126/science.aan4834.
13. Carlotta De Filippo et al., "Impact of Diet in Shaping Gut Microbiota Revealed by a Comparative Study in Children from Europe and Rural Africa,"

*Proceedings of the National Academy of Sciences* 107, no. 33 (Aug. 2010): 14691–6, doi: 10.1073/pnas.1005963107.

14. Geneviève Dubois et al., "The Inuit Gut Microbiome Is Dynamic Over Time and Shaped by Traditional Foods," *Microbiome* 5, no. 1 (Nov. 2017): 151, doi: 10.1186/s40168-017-0370-7, PubMed PMID: 29145891.
15. Pajau Vangay et al., "US Immigration Westernizes the Human Gut Microbiome," *Cell* 175, no. 4 (Nov. 2018): 962–72.e10, doi: 10.1016/j.cell.2018.10.029, PubMed PMID: 30388453.
16. Erica D. Sonnenburg and Justin L. Sonnenburg, "The Ancestral and Industrialized Gut Microbiota and Implications for Human Health," *Nature Reviews Microbiology* 17, no. 6 (June 2019): 383–90, doi: 10.1038/s41579-019-0191-8.
17. Maria Dominguez Bello et al.,"Preserving Microbial Diversity—Microbiota from Humans of All Cultures Are Needed to Ensure the Health of Future Generations," *Science* 362, no. 6410 (October 2018): 33–34.
18. Marta Selma-Royo et al., "Shaping Microbiota During the First 1,000 Days of Life," in: Stefano Guandalini and Flavia Indrio, eds., *Probiotics and Child Gastrointestinal Health*, Advances in Microbiology, Infectious Diseases and Public Health, vol. 10 (Cham, Switzerland: Springer International Publishing, 2019), 3–24.
19. Suma Magge and Anthony Lembo, "Low-FODMAP Diet for Treatment of Irritable Bowel Syndrome," *Gastroenterology & Hepatology* (NY) 8, no. 11 (Nov. 2012): 739–45, PubMed PMID: 24672410.
20. Karen L. Chen and Zeynep Madak Erdogan, "Estrogen and Microbiota Crosstalk: Should We Pay Attention?" *Trends in Endocrinology and Metabolism* 27, no. 11 (Aug. 2016): 752–55, doi: https://doi.org/10.1016/j.tem.2016.08.001.

Chapter Four: Stress and Brain Disorders

1. Andrea H. Weinberger et al., "Trends in Depression Prevalence in the USA from 2005 to 2015: Widening Disparities in Vulnerable Groups," *Psychological Medicine* 48, no. 8 (Oct. 2017): 1308–15, doi: 10.1017/S0033291717002781. Olle Hagnell et al., "Prevalence of Mental Disorders, Personality Traits and Mental Complaints in the Lundby Study: A Point Prevalence Study of the 1957 Lundby Cohort of 2,612 Inhabitants of a Geographically Defined Area Who Were Re-Examined in 1972 Regardless of Domicile," *Scandinavian Journal of Social Medicine Supplementum* 50 (1994): 1–77, doi: 10.2307/45199764.
2. Bruno Giacobbo et al., "Brain-Derived Neurotrophic Factor in Brain Disorders: Focus on Neuroinflammation," *Molecular Neurobiology* 56, no. 5 (May 2019): 3295–3312, doi: 10.1007/s12035-018-1283-6, PubMed PMID: 30117106.
3. Keenan A. Walker, "Inflammation and Neurodegeneration: Chronicity Matters," *Aging* (Albany, NY) 11, no. 1 (Dec. 2018): 3–4, doi: 10.18632/aging.101704, PubMed PMID: 30554190.

4. Huiying Wang et al., "*Bifidobacterium longum* 1714™ Strain Modulates Brain Activity of Healthy Volunteers During Social Stress," *American Journal of Gastroenterology* 114, no. 7 (July 2019): 1152–62.
5. Siddhartha Ghosh et al., "Intestinal Barrier Dysfunction, Lps Translocation and Disease Development," *Journal of the Endocrine Society* 4, no. 2 (February 2020): bvz039.
6. Pauline Luczynski et al., "Growing Up in a Bubble: Using Germ-Free Animals to Assess the Influence of the Gut Microbiota on Brain and Behavior," *International Journal of Neuropsychopharmacology* 19, no. 8 (Feb. 2016): pyw020, doi: 10.1093/ijnp/pyw020, PubMed PMID: 26912607.
7. Arthi Chinna Meyyappan et al., "Effect of Fecal Microbiota Transplant on Symptoms of Psychiatric Disorders: A Systematic Review," *BMC Psychiatry* 20, no. 1 (June 15, 2020): article 299, doi: 10.1186/s12888-020-02654-5.
8. Hai-yin Jiang et al., "Altered Fecal Microbiota Composition in Patients with Major Depressive Disorder," *Brain, Behavior, and Immunity* 48 (Aug. 2015): 186–94, doi: https://doi.org/10.1016/j.bbi.2015.03.016.
9. P. Zheng et al., "Gut Microbiome Remodeling Induces Depressive-like Behaviors Through a Pathway Mediated by the Host's Metabolism," *Molecular Psychiatry* 21, no. 6 (June 2016): 786–96, doi: 10.1038/mp.2016.44; and John Richard Kelly et al., "Transferring the Blues: Depression-Associated Gut Microbiota Induces Neurobehavioural Changes in the Rat," *Journal of Psychiatric Research* 82 (July 2016): 109–18, doi: https://doi.org/10.1016/j.jpsychires.2016.07.019.
10. Trisha A. Jenkins et al., "Influence of Tryptophan and Serotonin on Mood and Cognition with a Possible Role of the Gut-Brain Axis," *Nutrients* 8, no. 1 (Jan. 2016): 56, doi: 10.3390/nu8010056, PubMed PMID: 26805875.
11. Clair R. Martin et al., "The Brain-Gut-Microbiome Axis," *Cellular and Molecular Gastroenterology and Hepatology* 6, no. 2 (Apr. 2018): 133–48, doi: 10.1016/j.jcmgh.2018.04.003, PubMed PMID: 30023410.
12. Jessica M. Yano et al., "Indigenous Bacteria from the Gut Microbiota Regulate Host Serotonin Biosynthesis," *Cell* 161, no. 2 (Apr. 2015): 264–76, doi: 10.1016/j.cell.2015.02.047.
13. Thomas C. Fung et al., "Intestinal Serotonin and Fluoxetine Exposure Modulate Bacterial Colonization in the Gut," *Nature Microbiology* 4, no. 12 (Dec. 2019): 2064–73, doi: 10.1038/s41564-019-0540-4, PubMed PMID: 31477894.
14. Robert L. Stephens and Yvette Tache, "Intracisternal Injection of a TRH Analogue Stimulates Gastric Luminal Serotonin Release in Rats," *American Journal of Physiology: Gastrointestinal and Liver Physiology* 256, no. 2 (Feb. 1989): G377–G383, doi: 10.1152/ajpgi.1989.256.2.G377.
15. Vadim Osadchiy, Clair R. Martin, and Emeran A. Mayer, "Gut Microbiome and Modulation of CNS Function," *Comprehensive Physiology* 10, no. 1 (Dec. 18, 2019): 57–72, doi: doi:10.1002/cphy.c180031.

16. Ibid.
17. Iona A. Marin et al., "Microbiota Alteration Is Associated with the Development of Stress-Induced Despair Behavior," *Nature Scientific Reports* 7, no. 1 (Mar. 7, 2017): article 43859, doi: 10.1038/srep43859.
18. Vadim Osadchiy et al., "Correlation of Tryptophan Metabolites with Connectivity of Extended Central Reward Network in Healthy Subjects," *PloS One* 13, no. 8 (Aug. 2018): e0201772, doi: 10.1371/journal.pone.0201772, PubMed PMID: 30080865.
19. Christopher Brydges et al., for the Mood Disorders Precision Medicine Consortium, "Indoxyl Sulfate, a Gut Microbiome-Derived Uremic Toxin, Is Associated with Psychic Anxiety and Its Functional Magnetic Resonance Imaging-Based Neurologic Signature," doi: https://doi.org/10.1101/2020.12.08.388942.
20. Andrew C. Peterson and Chiang-Shan R. Li, "Noradrenergic Dysfunction in Alzheimer's and Parkinson's Diseases—An Overview of Imaging Studies," *Frontiers in Aging Neuroscience* 10 (May 1, 2018): article 127.
21. R. Alberto Travagli and Laura Anselmi, "Vagal Neurocircuitry and Its Influence on Gastric Motility," *Nature Reviews Gastroenterology & Hepatology* 13, no. 7 (May 2016): 389–401, doi: 10.1038/nrgastro.2016.76.
22. Andrée-Anne Poirier et al., "Gastrointestinal Dysfunctions in Parkinson's Disease: Symptoms and Treatments," *Parkinson's Disease* 2016, article 6762528, doi: 10.1155/2016/6762528.
23. Ibid.
24. Han-Lin Chiang and Chin-Hsien Lin, "Altered Gut Microbiome and Intestinal Pathology in Parkinson's Disease," *Journal of Movement Disorders* 12, no. 2 (May 2019): 67–83, doi: 10.14802/jmd.18067, PubMed PMID: 31158941.
25. Sara Gerhardt and Hasan Mohajeri, "Changes of Colonic Bacterial Composition in Parkinson's Disease and Other Neurodegenerative Diseases," *Nutrients* 10, no. 6 (June 2018): 708, doi: 10.3390/nu10060708.
26. Marcus M. Unger et al., "Short Chain Fatty Acids and Gut Microbiota Differ between Patients with Parkinson's Disease and Age-Matched Controls," *Parkinsonism & Related Disorders* 32 (Aug. 2016): 66–72, doi: https://doi.org/10.1016/j.parkreldis.2016.08.019.
27. Leo Galland, "The Gut Microbiome and the Brain," *Journal of Medicinal Food* 17, no. 12 (Nov. 2014): 1261–72, doi: 10.1089/jmf.2014.7000, PubMed PMID: 25402818.
28. Vayu Maini Rekdal et al., "Discovery and Inhibition of an Interspecies Gut Bacterial Pathway for Levodopa Metabolism," *Science* 364, no. 6445 (June 14, 2019): eaau6323, doi: 10.1126/science.aau6323.
29. Institute of Medicine, *Sleep Disorders and Sleep Deprivation: An Unmet Public Health Problem* (Washington, DC: National Academies Press, 2006).
30. Carlos H. Schenck, Bradley F. Boeve, and Mark W. Mahowald, "Delayed Emergence of a Parkinsonian Disorder in 38% of 29 Older Men Initially

Diagnosed with Idiopathic Rapid Eye Movement Sleep Behavior Disorder," *Neurology* 46, no. 2 (Feb. 1996): 388–93, doi: 10.1212/WNL.46.2.388.

31. Sadie Costello et al., "Parkinson's Disease and Residential Exposure to Maneb and Paraquat from Agricultural Applications in the Central Valley of California," *American Journal of Epidemiology* 169, no. 8 (Apr. 2009): 919–26, doi: 10.1093/aje/kwp006.
32. National Pesticide Information Center, "Diazinon," http://npic.orst.edu/factsheets/Diazgen.html.
33. Alzheimer's Association, "Alzheimer's Disease Facts and Figures," https://www.alz.org/alzheimers-dementia/facts-figures.
34. Judy George, "Gut-Liver-Brain Interactions Tied to Alzheimer's," July 26, 2018, https://www.medpagetoday.com/meetingcoverage/aaic/74246.
35. Kwangsik Nho et al., Alzheimer's Disease Neuroimaging I, the Alzheimer Disease Metabolomics C, "Altered Bile Acid Profile in Mild Cognitive Impairment and Alzheimer's Disease: Relationship to Neuroimaging and CSF Biomarkers," *Alzheimer's & Dementia* 15, no. 2 (Feb. 2019): 232–44, doi: 10.1016/j.jalz.2018.08.012, PubMed PMID: 30337152.
36. Matthew McMillin and Sharon DeMorrow, "Effects of Bile Acids on Neurological Function and Disease," *FASEB Journal* 30, no. 11 (Nov. 2016): 3658–68, doi: 10.1096/fj.201600275R.
37. Kwangsik Nho et al., "Altered Bile Acid Profile in Mild Cognitive Impairment and Alzheimer's Disease: Relationship to Neuroimaging and CSF Biomarkers," *Alzheimer's & Dementia* 15, no. 2 (February 2019): 232–244, doi: 10.1016/j.jalz.2018.08.012; Siamak Mahmoudian Dehkordi et al., "Altered Bile Acid Profile Associates with Cognitive Impairment in Alzheimer's Disease—An Emerging Role for Gut Microbiome," *Alzheimer's & Dementia* 15, no. 1 (January 2019): 76–92, doi: 10.1016/j.Jalz2018.07217.
38. Dianne Price, "Autism Symptoms Reduced Nearly 50% 2 Years after Fecal Transplant," Apr. 9, 2019, https://asunow.asu.edu/20190409-discoveries-autism-symptoms-reduced-nearly-50-percent-two-years-after-fecal-transplant.
39. David Q. Beversdorf, Hanna E. Stevens, and Karen L. Jones, "Prenatal Stress, Maternal Immune Dysregulation, and Their Association with Autism Spectrum Disorders," *Current Psychiatry Reports* 20, no. 9 (Aug. 2018): article 76, doi: 10.1007/s11920-018-0945-4, PubMed PMID: 30094645.
40. Helen E. Vuong and Elaine Y. Hsiao, "Emerging Roles for the Gut Microbiome in Autism Spectrum Disorder," *Biological Psychiatry* 81, no. 5 (Mar. 1, 2017): 411–23, doi: 10.1016/j.biopsych.2016.08.024, PubMed PMID: 27773355.
41. Katherine M. Flegal et al., "Prevalence and Trends in Obesity among US Adults, 1999–2008," *Journal of the American Medical Association* 303, no. 3 (Jan. 2010): 235–41, doi: 10.1001/jama.2009.2014; and R. Bethene Ervin, "Prevalence of Metabolic Syndrome among Adults 20 Years of Age and Over, by Sex, Age, Race and Ethnicity, and Body Mass Index: United States, 2003–

2006," National Health Statistics Reports, no. 13 (2009): 1–7, PubMed PMID: 19634296.

42. Rosa Krajmalnik-Brown et al., "Gut Bacteria in Children with Autism Spectrum Disorders: Challenges and Promise of Studying How a Complex Community Influences a Complex Disease," *Microbial Ecology in Health and Disease* 26 (Mar. 2015): article 26914, doi: 10.3402/mehd.v26.26914, PubMed PMID: 25769266.
43. Dae-Wook Kang et al., "Reduced Incidence of *Prevotella* and Other Fermenters in Intestinal Microflora of Autistic Children," *PLoS One* 8, no. 7 (July 3, 2013): e68322, doi: 10.1371/journal.pone.0068322.
44. Dae-Wook Kang et al., "Microbiota Transfer Therapy Alters Gut Ecosystem and Improves Gastrointestinal and Autism Symptoms: An Open-Label Study," *Microbiome* 5, no. 1 (Jan. 2017): 10, doi: 10.1186/s40168-016-0225-7.
45. Dae-Wook Kang et al., "Long-Term Benefit of Microbiota Transfer Therapy on Autism Symptoms and Gut Microbiota," *Nature Scientific Reports* 9, no. 1 (Apr. 2019): 5821, doi: 10.1038/s41598-019-42183-0.
46. Dianne Price, "Autism Symptoms Reduced Nearly 50% 2 Years after Fecal Transplant," *ASU News*, April 9, 2019, https://asunow.asu.edu/20190409-discoveries-autism-symptoms-reduced-nearly-50-percent-two-years-after-fecal-transplant.
47. Kate Julian, "What Happened to American Childhood?" *Atlantic*, May 2020, https://www.theatlantic.com/magazine/archive/2020/05/childhood-in-an-anxious-age/609079.
48. Iona A. Marin et al., "Microbiota Alteration Is Associated with the Development of Stress-Induced Despair Behavior," *Nature Scientific Reports* 7, no. 1 (Mar. 7, 2017): article 43859, doi: 10.1038/srep43859.

## Chapter Five: How Diet Regulates the Brain-Gut-Microbiome Network

1. Isabella Meira et al., "Ketogenic Diet and Epilepsy: What We Know So Far," *Frontiers in Neuroscience* 13 (Jan. 2019): article 5, doi: 10.3389/fnins.2019.00005, PubMed PMID: 30760973.
2. Martin Kohlmeier, *Nutrient Metabolism: Structures, Functions, and Genes*, 2nd ed. (Cambridge, MA: Academic Press, 2015), 111–86.
3. Christine Olson, Helen Vuong, and Jessica M. Yano, "The Gut Microbiota Mediates the Anti-Seizure Effects of the Ketogenic Diet," *Cell* 173, no. 7 (May 2018): 1728–41.e13, doi: 10.1016/j.cell.2018.04.027.
4. Victoria M. Gershuni, Stephanie L. Yan, and Valentina Medici, "Nutritional Ketosis for Weight Management and Reversal of Metabolic Syndrome," *Current Nutrition Reports* 7, no. 3 (Sept. 2018): 97–106, doi: 10.1007/s13668-018-0235-0.
5. O. Henríquez Sánchez et al., "Adherence to the Mediterranean Diet and Quality of Life in the SUN Project," *European Journal of Clinical Nutrition* 66, no. 3 (Mar. 2012): 360–68, doi: 10.1038/ejcn.2011.146.
6. Maria Shadrina, Elena A. Bondarenko, and Petr A. Slominsky, "Genetics

Factors in Major Depression Disease," *Frontiers in Psychiatry* (Sept. 2018): 334, doi: 10.3389/fpsyt.2018.00334.

7. Marc Molendijk et al., "Diet Quality and Depression Risk: A Systematic Review and Dose-Response Meta-Analysis of Prospective Studies," *Journal of Affective Disorders* 226 (Jan. 15, 2018): 346–54, doi: 10.1016/j.jad.2017.09.022.
8. Theodora Psaltopoulou et al., "Mediterranean Diet, Stroke, Cognitive Impairment, and Depression: A Meta-Analysis," *Annals of Neurology* 74, no. 4 (Oct. 2013): 580–91, doi: 10.1002/ana.23944.
9. Almudena Sánchez-Villegas and Ana Sánchez-Tainta, *The Prevention of Cardiovascular Disease through the Mediterranean Diet*, 1st ed. (Cambridge, MA: Academic Press, 2017).
10. Natalie Parletta et al., "A Mediterranean-Style Dietary Intervention Supplemented with Fish Oil Improves Diet Quality and Mental Health in People with Depression: A Randomized Controlled Trial (HELFIMED)," *Nutritional Neuroscience* 22, no. 1 (Dec. 2017): 1–14, doi: 10.1080/1028415X.2017.1411320.
11. Felice N. Jacka et al., "A Randomised Controlled Trial of Dietary Improvement for Adults with Major Depression (the 'SMILES' Trial)," *BMC Medicine* 15, no. 1 (Jan. 30, 2017): article 23, doi: 10.1186/s12916-017-0791-y.
12. Food and Mood Centre, The SMILEs Trial, https://foodandmoodcentre.com.au/smiles-trial.
13. Paola Vitaglione et al., "Biomarkers of Intake of a Mediterranean Diet: Which Contribution from the Gut Microbiota?" *Nutrition, Metabolism and Cardiovascular Diseases* 29, no. 8 (Aug. 2019): 880, doi: 10.1016/j.numecd.2019.05.034.
14. Scott C. Anderson, John F. Cryan, and Ted Dinan, *The Psychobiotic Revolution: Mood, Food, and the New Science of the Gut-Brain Connection* (Washington, DC: National Geographic, 2017).
15. Asma Kazemi et al., "Effect of Probiotic and Prebiotic vs Placebo on Psychological Outcomes in Patients with Major Depressive Disorder: A Randomized Clinical Trial," *Clinical Nutrition* (Edinburgh) 38, no. 2 (Apr. 2019): 522–28, doi: 10.1016/j.clnu.2018.04.010, PubMed PMID: 29731182.
16. R. F. Slykerman et al., "Effect of *Lactobacillus rhamnosus* HN001 in Pregnancy on Postpartum Symptoms of Depression and Anxiety: A Randomised Double-Blind Placebo-Controlled Trial," *EBioMedicine* 24C (Sept. 2017): 159–65, doi: 10.1016/j.ebiom.2017.09.013.
17. Amory Meltzer and Judy Van de Water, "The Role of the Immune System in Autism Spectrum Disorder," *Neuropsychopharmacology* 42, no. 1 (Jan. 2017): 284–98, doi: 10.1038/npp.2016.158.
18. Charlotte Madore et al., "Neuroinflammation in Autism: Plausible Role of Maternal Inflammation, Dietary Omega 3, and Microbiota," *Neural Plasticity* 2016, no. 3: 1–15, doi: 10.1155/2016/3597209, PubMed PMID: 27840741.
19. Jun Ma et al., "High-Fat Maternal Diet During Pregnancy Persistently Alters the Offspring Microbiome in a Primate Model," *Nature Communications* 5, no. 1 (2014): article 3889, doi: 10.1038/ncomms4889.

20. Shelly A. Buffington et al., "Microbial Reconstitution Reverses Maternal Diet-Induced Social and Synaptic Deficits in Offspring," *Cell* 165, no. 7 (June 2016): 1762–75, doi: 10.1016/j.cell.2016.06.001, PubMed PMID: 27315483.
21. Richard H. Sandler et al., "Short-Term Benefit from Oral Vancomycin Treatment of Regressive-Onset Autism," *Journal of Child Neurology* 15, no. 7 (Aug. 2000): 429–35, doi: 10.1177/088307380001500701.
22. Felice N. Jacka et al., "Western Diet Is Associated with a Smaller Hippocampus: A Longitudinal Investigation," *BMC Medicine* 13, no. 1 (Sept. 2015): article 215, doi: 10.1186/s12916-015-0461-x, PubMed PMID: 26349802.
23. National Heart, Lung, and Blood Institute, "DASH Eating Plan," https://www.nhlbi.nih.gov/health-topics/dash-eating-plan.
24. Martha Clare Morris et al., "MIND Diet Slows Cognitive Decline with Aging," *Alzheimer's & Dementia* 11, no. 9 (Sept. 2015): 1015–22, doi: 10.1016/j.jalz.2015.04.011.
25. Marta Grochowska, Tomasz Laskus, and Marek Radkowski, "Gut Microbiota in Neurological Disorders," *Archivum Immunologiae et Therapiae Experimentalis* 67, no. 6 (Oct. 2019): 375–83, doi: 10.1007/s00005-019-00561-6.
26. Rasnik K. Singh et al., "Influence of Diet on the Gut Microbiome and Implications for Human Health," *Journal of Translational Medicine* 15, no. 1 (Apr. 8, 2017): article 73, doi: 10.1186/s12967-017-1175-y, PubMed PMID: 28388917.
27. Tarini Shankar Ghosh et al., "Mediterranean Diet Intervention Alters the Gut Microbiome in Older People Reducing Frailty and Improving Health Status: The NU-AGE 1-Year Dietary Intervention Across Five European Countries," *Gut* 67, no. 7 (Feb. 2020): 1218–28, doi: 10.1136/gutjnl-2019-319654.
28. Siamak Mahmoudiandehkordi et al., "Altered Bile Acid Profile Associates with Cognitive Impairment in Alzheimer's Disease—an Emerging Role for Gut Microbiome," *Alzheimer's & Dementia* 15, no. 1 (Oct. 2019): 76–92, doi: 10.1016/j.jalz.2018.07.217, PubMed PMID: 30337151.

Chapter Six: A Broader Connection:
How Exercise and Sleep Affect Our Microbiome

1. Yanping Li et al., "Healthy Lifestyle and Life Expectancy Free of Cancer, Cardiovascular Disease, and Type 2 Diabetes: Prospective Cohort Study," *British Medical Journal* 368 (Jan. 2020): article l6669, doi: 10.1136/bmj.l6669, PubMed PMID: 31915124.
2. Solja T. Nyberg et al., "Association of Healthy Lifestyle with Years Lived without Major Chronic Diseases," *JAMA Internal Medicine* 180, no. 5 (May 2020): 1–10, doi: 10.1001/jamainternmed.2020.0618, PubMed PMID: 32250383.
3. Cassie M. Mitchell et al., "Does Exercise Alter Gut Microbial Composition? A Systematic Review," *Medicine and Science in Sports and Exercise* 51, no. 1 (Aug. 2018), 160–67, doi: 10.1249/MSS.0000000000001760.
4. Siobhan F. Clarke et al., "Exercise and Associated Dietary Extremes Im-

pact on Gut Microbial Diversity," *Gut* 63, no. 12 (Dec. 2014): 1913–20, doi: 10.1136/gutjnl-2013-306541, PubMed PMID: 25021423.

5. Jacob Allen et al., "Exercise Alters Gut Microbiota Composition and Function in Lean and Obese Humans," *Medicine & Science in Sports & Exercise* 50, no. 4 (Apr. 2018): 747–57.
6. J. Philip Karl et al., "Changes in Intestinal Microbiota Composition and Metabolism Coincide with Increased Intestinal Permeability in Young Adults under Prolonged Physiological Stress," *American Journal of Physiology—Gastrointestinal and Liver Physiology* 312, no. 6 (June 2017): G559–G571, doi: 10.1152/ajpgi.00066.2017.
7. Núria Mach and Dolors Fuster-Botella, "Endurance Exercise and Gut Microbiota: A Review," *Journal of Sport and Health Science* 6, no. 2 (May 2017): 179–97, doi: 10.1016/j.jshs.2016.05.001, PubMed PMID: 30356594.
8. Erick Prado de Oliveira, Roberto Carlos Burini, and Asker Jeukendrup, "Gastrointestinal Complaints During Exercise: Prevalence, Etiology, and Nutritional Recommendations," *Sports Medicine* 44, suppl.1 (2014): S79–S85, doi: 10.1007/s40279-014-0153-2, PubMed PMID: 24791919.
9. David Ferry, "Does Your Gut Hold the Secret to Performance?" Outside, Jan. 15, 2018, https://www.outsideonline.com/2274441/no-gut-no-glory.
10. Jonathan Scheiman et al., "Meta-omics Analysis of Elite Athletes Identifies a Performance-Enhancing Microbe That Functions via Lactate Metabolism," *Nature Medicine* 25, no. 7 (July 2019): 1104–9, doi: 10.1038/s41591-019-0485-4.
11. Abiola Keller et al., "Does the Perception That Stress Affects Health Matter? The Association with Health and Mortality," *Health Psychology* 31, no. 5 (Sept. 2012): 677–84, doi: 10.1037/a0026743; and Kari Leibowitz and Alia Crum, "In Stressful Times, Make Stress Work for You," *New York Times*, Apr. 1, 2020.
12. Alana Conner et al., "Americans' Health Mindsets: Content, Cultural Patterning, and Associations with Physical and Mental Health," *Annals of Behavioral Medicine* 53, no. 4 (June 2018): 321–32, doi: 10.1093/abm/kay041.
13. Michael Pollan, "Our National Eating Disorder," *New York Times Magazine*, Oct. 17, 2004, https://www.nytimes.com/2004/10/17/magazine/our-national-eating-disorder.html.
14. Paul N. Rozin et al., "Attitudes to Food and the Role of Food in Life in the USA, Japan, Flemish Belgium, and France: Possible Implications for the Diet-Health Debate," *Appetite* 33, no. 2 (Oct. 1999): 163–80, doi: https://doi.org/10.1006/appe.1999.0244.
15. Kaitlin Woolley and Ayelet Fishbach, "For the Fun of It: Harnessing Immediate Rewards to Increase Persistence in Long-Term Goals," *Journal of Consumer Research* 42, no. 6 (Apr. 2016): 952–66, doi: 10.1093/jcr/ucv098.
16. Bradley P. Turnwald et al., "Increasing Vegetable Intake by Emphasizing Tasty and Enjoyable Attributes: A Randomized Controlled Multisite Intervention

for Taste-Focused Labeling," *Psychological Science* 30, no. 11 (Nov. 2019): 1603–15, doi: 10.1177/0956797619872191.

17. Bradley P. Turnwald, Danielle Z. Boles, and Alia J. Crum, "Association between Indulgent Descriptions and Vegetable Consumption: Twisted Carrots and Dynamite Beets," *JAMA Internal Medicine* 177, no. 8 (Aug. 2017): 1216–18, doi: 10.1001/jamainternmed.2017.1637.
18. Luciana Besedovsky, Tanja Lange, and Monika Haack, "The Sleep-Immune Crosstalk in Health and Disease," *Physiological Reviews* 99, no. 3 (July 1, 2019): 1325–80, doi: 10.1152/physrev.00010.2018, PubMed PMID: 30920354.
19. Christoph A. Thaiss et al., "Transkingdom Control of Microbiota Diurnal Oscillations Promotes Metabolic Homeostasis," *Cell* 159, no. 3 (Oct. 2014): 514–29, doi: 10.1016/j.cell.2014.09.048.

Chapter Seven: Restoring the Gut Microbiome

1. Chana Davis, "How Much Protein Do I Need?" https://medium.com/@chanapdavis/how-much-protein-do-you-need-37143cb0d499.
2. Jiaqui Huang et al., "Association between Plant and Animal Protein Intake and Overall and Cause-Specific Mortality," *JAMA Internal Medicine* 180, no. 9 (Sept. 1, 2020): 1173–84, doi: 10.1001/jamainternmed.2020.2790.
3. Katríona E. Lyons et al., "Breast Milk, a Source of Beneficial Microbes and Associated Benefits for Infant Health," *Nutrients* 12, no. 4 (Apr. 2020): article 1039, doi: 10.3390/nu12041039, PubMed PMID: 32283875.
4. Šárka Musilová et al., "Beneficial Effects of Human Milk Oligosaccharides on Gut Microbiota," *Beneficial Microbes* 5, no. 3 (Sept. 2014): 273–83, doi: 10.3920/bm2013.0080, PubMed PMID: 24913838.
5. Michael Pollan, "Some of My Best Friends Are Germs," *New York Times Magazine*, May 15, 2013, https://www.nytimes.com/2013/05/19/magazine/say-hello-to-the-100-trillion-bacteria-that-make-up-your-microbiome.html.
6. Long Ge et al., "Comparison of Dietary Macronutrient Patterns of 14 Popular Named Dietary Programmes for Weight and Cardiovascular Risk Factor Reduction In Adults: Systematic Review and Network Meta-Analysis of Randomised Trials," *British Medical Journal* 369 (2020): m696, doi: 10.1136/bmj.m696.
7. John B. Furness and David M. Bravo, "Humans as Cucinivores: Comparisons with Other Species," *Journal of Comparative Physiology B* 185, no. 8 (Dec. 2015): 1–10, doi: 10.1007/s00360-015-0919-3.
8. D. Rosenberg and F. Klimscha, "Prehistoric Dining at Tel Tsaf," *Biblical Archaeological Review* 44, no. 4 (July/August 2018).
9. Ibid.
10. Erica D. Sonnenburg and Justin L. Sonnenburg, "Starving Our Microbial Self: The Deleterious Consequences of a Diet Deficient in Microbiota-Accessible Carbohydrates," *Cell Metabolism* 20, no. 5 (Aug. 2014): 779–86, doi: 10.1016/j.cmet.2014.07.003, PubMed PMID: 25156449.
11. David Klurfeld et al., "Considerations for Best Practices in Studies of Fiber or

Other Dietary Components and the Intestinal Microbiome," *American Journal of Physiology—Endocrinology and Metabolism* 315, no. 6 (Aug. 2018): E1087–E1097, doi: 10.1152/ajpendo.00058.2018.

12. Boushra Dalile et al., "The Role of Short-Chain Fatty Acids in Microbiota-Gut-Brain Communication," *Nature Reviews Gastroenterology & Hepatology* 16, no. 8 (Aug. 2019): 461–78, doi: 10.1038/s41575-019-0157-3.
13. Sonnenburg and Sonnenburg, "Starving Our Microbial Self."
14. Denis P. Burkitt, Alec R. P. Walker, and Neil S. Painter, "Effect of Dietary Fibre on Stools and Transit Times, and Its Role in the Causation of Disease," *Lancet* 300, no. 7792 (Dec. 30, 1972): 1408–11, doi: 10.1016/S0140-6736(72)92974-1.
15. Sonnenburg and Sonnenburg, "Starving Our Microbial Self."
16. Jan-Hendrik Hehemann et al., "Bacteria of the Human Gut Microbiome Catabolize Red Seaweed Glycans with Carbohydrate-Active Enzyme Updates from Extrinsic Microbes," *Proceedings of the National Academy of Sciences* 109, no. 48 (Nov. 2012): 19786–91, doi: 10.1073/pnas.1211002109, PubMed PMID: 23150581.
17. Sonnenburg and Sonnenburg, "Starving Our Microbial Self."
18. Boushra Dalile et al., "The Role of Short-Chain Fatty Acids in Microbiota-Gut-Brain Communication," *Nature Review Gastroenterology & Hepatology* 16, no. 8 (Aug. 2019): 461–78; Erica Sonnenburg and Justin Sonnenburg, "Starving Our Microbial Self: The Deleterious Consequences of a Diet Deficient in Microbiota-Accessible Carbohydrates," *Cell Metabolism* 20, no. 5 (November 2014): 779–786.
19. Fernando Cardona et al., "Benefits of Polyphenols on Gut Microbiota and Implications in Human Health," *Journal of Nutritional Biochemistry* 24, no. 8 (Aug. 2013): 1415–22, doi: https://doi.org/10.1016/j.jnutbio.2013.05.001.
20. Senem Kamiloglu et al., "Anthocyanin Absorption and Metabolism by Human Intestinal Caco-2 Cells: A Review," *International Journal of Molecular Science* 16, no. 9 (Sept. 8, 2015): 21555–74, doi: 10.3390/ijms160921555, PubMed PMID: 26370977.
21. Colin D. Kay et al., "Anthocyanins and Flavanones Are More Bioavailable Than Previously Perceived: A Review of Recent Evidence," *Annual Review of Food Science and Technology* 8, no. 1 (Feb. 28, 2017): 155–80, doi: 10.1146/annurev-food-030216-025636.
22. Dagfinn Aune et al., "Fruit and Vegetable Intake and the Risk of Cardiovascular Disease, Total Cancer, and All-Cause Mortality: A Systematic Review and Dose-Response Meta-Analysis of Prospective Studies," *International Journal of Epidemiology* 46, no. 3 (June 1, 2017): 1029–56, doi: 10.1093/ije/dyw319, PubMed PMID: 28338764.
23. Ke Shen, Bin Zhang, and Qiushi Feng, "Association between Tea Consumption and Depressive Symptom among Chinese Older Adults," *BMC Geriatrics* 19, no. 1 (Sept. 2019): article 246, doi: 10.1186/s12877-019-1259-z.

24. Louise Hartley et al., "Green and Black Tea for the Primary Prevention of Cardiovascular Disease," *Cochrane Database of Systematic Reviews* 6, no. 6 (June 2013): article CD009934, doi: 10.1002/14651858.CD009934.pub2, PubMed PMID: CD009934; Shinichi Kuriyama, "The Relation between Green Tea Consumption and Cardiovascular Disease as Evidenced by Epidemiological Studies," *Journal of Nutrition* 138, no. 8 (Aug. 2008): 1548S–1553S, doi: 10.1093/jn/138.8.1548S; Taichi Shimazu et al., "Dietary Patterns and Cardiovascular Disease Mortality in Japan: A Prospective Cohort Study," *International Journal of Epidemiology* 36, no. 3 (June 2007): 600–609, doi: 10.1093/ije/dym005; and P. Elliott Miller et al., "Associations of Coffee, Tea, and Caffeine Intake with Coronary Artery Calcification and Cardiovascular Events," *American Journal of Medicine* 130, no. 2 (Feb. 2017): 188–97, doi: 10.1016/j.amjmed.2016.08.038, PubMed PMID: 27640739.
25. Sabu M. Chacko et al., "Beneficial Effects of Green Tea: A Literature Review," *Chinese Medicine* 5, no. 1 (Apr. 2010): 13, doi: 10.1186/1749-8546-5-13, PubMed PMID: 20370896.
26. Shen, Zhang, and Feng, "Association between Tea Consumption and Depressive Symptom."
27. Naghma Khan and Hasan Mukhtar, "Tea Polyphenols for Health Promotion," *Life Sciences* 81, no. 7 (Aug. 2007): 519–33, doi: 10.1016/j.lfs.2007.06.011, PubMed PMID: 17655876.
28. Fei-Yan Fan, Li-Xuan Sang, and Min Jiang, "Catechins and Their Therapeutic Benefits to Inflammatory Bowel Disease," *Molecules* 22, no. 3 (Mar. 19, 2017): 484, doi: 10.3390/molecules22030484, PubMed PMID: 28335502.
29. Carolina Cueva et al., "An Integrated View of the Effects of Wine Polyphenols and Their Relevant Metabolites on Gut and Host Health," *Molecules* 22, no. 1 (Jan. 6, 2017): 99, doi: 10.3390/molecules22010099, PubMed PMID: 28067835.
30. Caroline Le Roy et al., "Red Wine Consumption Associated with Increased Gut Microbiota α-Diversity in 3 Independent Cohorts," *Gastroenterology* 158, no. 1 (Aug. 2020): 270-272.e2, doi: 10.1053/j.gastro.2019.08.024.
31. Ibid.
32. Alexander Yashin et al., "Antioxidant Activity of Spices and Their Impact on Human Health: A Review," *Antioxidants* (Basel) 6, no. 3 (Sept. 2017): 70, doi: 10.3390/antiox6030070.
33. Nassima Talhaoui et al., "From Olive Fruits to Olive Oil: Phenolic Compound Transfer in Six Different Olive Cultivars Grown under the Same Agronomical Conditions," *International Journal of Molecular Science* 17, no. 3 (Mar. 2016): 337, doi: 10.3390/ijms17030337, PubMed PMID: 26959010.
34. Lara Costantini et al., "Impact of Omega-3 Fatty Acids on the Gut Microbiota," *International Journal of Molecular Science* 18, no. 12 (Dec. 2017): 2645, doi: 10.3390/ijms18122645; and Henry Watson et al., "A Randomised Trial of the Effect of Omega-3 Polyunsaturated Fatty Acid Supplements on the Hu-

man Intestinal Microbiota," *Gut* 67, no. 11 (Nov. 2018): 1974–83, doi: 10.1136/gutjnl-2017-314968.

35. Ruth E. Patterson et al., "Intermittent Fasting and Human Metabolic Health," *Journal of the American Academy of Nutrition and Dietetics* 115, no. 8 (Apr. 2015): 1203–12, doi: 10.1016/j.jand.2015.02.018, PubMed PMID: 25857868.
36. Francesco Sofi, "Fasting-Mimicking Diet: A Clarion Call for Human Nutrition Research or an Additional Swan Song for a Commercial Diet?" *International Journal of Food Sciences and Nutrition* 71, no. 8 (Dec. 2020): 921–28, doi: 10.1080/09637486.2020.1746959.
37. Leanne Harris et al., "Intermittent Fasting Interventions for Treatment of Overweight and Obesity in Adults: A Systematic Review and Meta-Analysis," *JBI Database of Systematic Reviews and Implementation Reports* 16, no. 2 (Feb. 2018): 507–47.
38. Vanessa Leone et al., "Effects of Diurnal Variation of Gut Microbes and High-Fat Feeding on Host Circadian Clock Function and Metabolism," *Cell Host & Microbe* 17, no. 5 (Apr. 2015): 681–89, doi: 10.1016/j.chom.2015.03.006, PubMed PMID: 25891358.
39. Amandine Chaix et al., "Time-Restricted Eating to Prevent and Manage Chronic Metabolic Diseases," *Annual Review of Nutrition* 39, no. 1 (Aug. 2019): 291–315, doi: 10.1146/annurev-nutr-082018-124320.
40. Amandine Chaix and Amir Zarrinpar, "The Effects of Time-Restricted Feeding on Lipid Metabolism and Adiposity," *Adipocyte* 4, no. 4 (May 2015): 319–24, doi: 10.1080/21623945.2015.1025184, PubMed PMID: 26451290.
41. Dylan A. Lowe et al., "Effects of Time-Restricted Eating on Weight Loss and Other Metabolic Parameters in Women and Men with Overweight and Obesity: The TREAT Randomized Clinical Trial," *JAMA Internal Medicine* 180, no. 11 (Sept. 28, 2020): 1491–99, doi: 10.1001/jamainternmed.2020.4153.

Chapter Eight: The Key to Gut Health Is in the Soil

1. Peter Bakker et al., "The Rhizosphere Revisited: Root Microbiomics," *Frontiers in Plant Science* 4 (May 30, 2013): 165; Roeland L. Berendsen, Corné Pieterse, and Peter Bakker, "The Rhizosphere Microbiome and Plant Health," *Trends in Plant Science* 17, no. 8 (May 2012): 478–86, doi: https://doi.org/10.1016/j.tplants.2012.04.001; and Stéphane Hacquard et al., "Microbiota and Host Nutrition across Plant and Animal Kingdoms," *Cell Host & Microbe* 17, no. 5 (May 2015): 603–16, doi: https://doi.org/10.1016/j.chom.2015.04.009.
2. Shamayim T. Ramírez-Puebla et al., "Gut and Root Microbiota Commonalities," *Applied and Environmental Microbiology* 79, no. 1 (Jan. 2013): 2–9, doi: 10.1128/AEM.02553-12.
3. Noah Fierer et al., "Reconstructing the Microbial Diversity and Function of Pre-Agricultural Tallgrass Prairie Soils in the United States," *Science* 342, no. 6158 (Nov. 1, 2013): 621–24, doi: 10.1126/science.1243768.

4. Landscope America, "Tallgrass Prairie Ecosystem," http://www.landscope.org/explore/ecosystems/disappearing_landscapes/tallgrass_prairie.
5. David R. Montgomery and Anne Biklé, *The Hidden Half of Nature: The Microbial Roots of Life and Health*, 1st ed. (New York: Norton, 2015).
6. Kishan Mahmud et al., "Current Progress in Nitrogen Fixing Plants and Microbiome Research," *Plants* 9, no. 1 (Jan. 2020): 97, https://doi.org/10.3390/plants9010097.
7. Alyson E. Mitchell et al., "Ten-Year Comparison of the Influence of Organic and Conventional Crop-Management Practices on the Content of Flavonoids in Tomatoes," *Journal of Agricultural and Food Chemistry* 55, no. 15 (July 2007): 6154–59, doi: 10.1021/jf070344+.
8. Richard Jacoby et al., "The Role of Soil Microorganisms in Plant Mineral Nutrition: Current Knowledge and Future Directions," *Frontiers in Plant Science* 8 (Sept. 19, 2017): 1617.
9. Jeyasankar Alagarmalai, "Phytochemicals: As Alternate to Chemical Pesticides for Insects Pest Management," *Current Trends Biomedical Engineering & Biosciences* 4, no. 1 (May 2017): 3–4, doi: 10.19080/CTBEB.2017.04.555627.
10. Marc-André Selosse, Alain Bessis, and María J. Pozo, "Microbial Priming of Plant and Animal Immunity: Symbionts as Developmental Signals," *Trends in Microbiology* 22, no. 11 (Nov. 2014): 607–13, doi: https://doi.org/10.1016/j.tim.2014.07.003.
11. Jing Gao et al., "Impact of the Gut Microbiota on Intestinal Immunity Mediated by Tryptophan Metabolism," *Frontiers in Cellular and Infection Microbiology* 8 (2018): 13; and Jessica M. Yano et al., "Indigenous Bacteria from the Gut Microbiota Regulate Host Serotonin Biosynthesis," *Cell* 161, no. 2 (Apr. 9, 2015): P264–P276, doi: 10.1016/j.cell.2015.02.047.
12. Vadim Osadchiy et al., "Correlation of Tryptophan Metabolites with Connectivity of Extended Central Reward Network in Healthy Subjects, *PloS One* 13, no. 8 (Aug. 6, 2018): e0201772-e, doi: 10.1371/journal.pone.0201772, PubMed PMID: 30080865.
13. Wikipedia, "Justus von Liebig," https://en.wikipedia.org/wiki/Justus_von_Liebig; and Margaret W. Rossiter, *The Emergence of Agricultural Science: Justus Liebig and the Americans, 1840–1880* (New Haven, CT: Yale University Press, 1975).
14. Wikipedia, "Green Revolution," https://en.wikipedia.org/wiki/Green_Revolution; and Hari Krishan Jain, *The Green Revolution: History, Impact and Future* (Houston, Studium Press, 2010).
15. David R. Montgomery and Anne Biklé, *The Hidden Half of Nature: The Microbial Roots of Life and Health*, 1st ed. (New York: Norton, 2015).
16. Anne Biklé and David R. Montgomery, "Junk Food Is Bad for Plants, Too," Nautilus, Mar. 31, 2016, http://nautil.us/issue/34/adaptation/junk-food-is-bad-for-plants-too.

17. Ibid.
18. US Environmental Protection Agency, "Organic Farming," https://www.epa.gov/agriculture/organic-farming.
19. Miles McEvoy, "Organic 101: What the USDA Organic Label Means," US Department of Agriculture, https://www.usda.gov/media/blog/2012/03/22/organic-101-what-usda-organic-label-means.
20. Rodale Institute, "Regenerative Organic Agriculture and Climate Change: A Down-to-Earth Solution to Global Warming," 2014, https://rodaleinstitute.org/wp-content/uploads/rodale-white-paper.pdf.
21. Ibid.


## Chapter Nine: The One-Health Concept


1. Bin Ma et al., "Earth Microbial Co-occurrence Network Reveals Interconnection Pattern across Microbiomes," *Microbiome* 8, 82 (2020), doi: 10.1186/s40168-020-00857-2.
2. Earth Microbiome Project, https://earthmicrobiome.org.
3. World Economic Forum, "Save the Axolotl: Dangers of Accelerated Biodiversity Loss," https://reports.weforum.org/global-risks-report-2020/save-the-axolotl; and Eric Chivian and Aaron Bernstein, eds., *Sustaining Life: How Human Health Depends on Biodiversity* (New York: Oxford University Press, 2008).
4. Delphine Destoumieux-Garzón et al., "The One-Health Concept: Ten Years Old and a Long Road Ahead," *Frontiers in Veterinary Science* 5 (Feb. 2018): 14.
5. Ibid.
6. Walter Willett et al., "Food in the Anthropocene: The EAT–Lancet Commission on Healthy Diets from Sustainable Food Systems," *Lancet* 393, no. 10170 (Feb. 2, 2019): 447–92, doi: 10.1016/S0140-6736(18)31788-4.
7. Frank B. Hu, Brett O. Otis, Gina McCarthy, "Can Plant-Based Meat Alternatives Be Part of a Healthy and Sustainable Diet?," *JAMA*, 2019; 322(16):1547–1548.
8. Anahad O'Connor, "Fake Meat vs. Real Meat," *New York Times*, December 3, 2019, nytimes.com/2019/12/03/well/eat/fake-meat-vs-real-meat.html?smid=em-share
9. Yvon Chouinard, "Why Food?" Patagonia Provisions, Apr. 23, 2020, https://www.patagoniaprovisions.com/pages/why-food-essay.
10. Ibid.
11. Yvon Chouinard, *Let My People Go Surfing: The Education of a Reluctant Businessman* (New York: Penguin Books, 2006).
12. Patagonia Provisions, "B Lab," B Corp statement, https://www.patagonia.com/b-lab.html.
13. Certified B Corporation, https://bcorporation.net.
14. Danone, "B Corp," https://www.danone.com/about-danone/sustainable-value-creation/BCorpAmbition.html.

15. Ann Abel, "Local, Sustainable, and Delicious: Here's How the Coronavirus Helped One Michelin Chef Share His Food Philosophy," *Forbes*, June 16, 2020, https://www.forbes.com/sites/annabel/2020/06/16/local-sustainable-and-delicious-heres-how-the-coronavirus-helped-one-michelin-chef-share-his-food-philosophy.

# INDEX

Page numbers followed by n indicate notes.

# ABOUT THE AUTHOR

Emeran Mayer, MD, is the author of *The Mind-Gut Connection*. He has studied brain-body interactions for forty years and is the executive director of the G. Oppenheimer Center for Neurobiology of Stress and Resilience and the director of the UCLA Microbiome Center. His research has been supported by the National Institutes of Health for the past twenty-five years, and he is considered a pioneer and world leader in the area of brain-gut-microbiome interactions.